NICOTINE

by Jenny Rackley

DRUG
EDUCATION
LIBRARY

LUCENT BOOKS
SAN DIEGO, CALIFORNIA

THOMSON
━━━━━✦━━━━━ ™
GALE

Detroit • New York • San Diego • San Francisco
Boston • New Haven, Conn. • Waterville, Maine
London • Munich

Picture Credits:

Cover photo: © Gary W. Carter/CORBIS
© Kirk Anderson, 83
Associated Press, 38
Associated Press/Phil Coale, 56
Bennett © North America Syndicate, 42
© Bettmann/CORBIS, 14, 26, 78
Brookins © *Richmond Times-Dispatch*, 32
© Hulton/Archive by Getty Images, 16, 24, 75, 77, 82
© Steve Kelley, 37
Liaison, 85
Library of Congress, 12
National Archives, 23
© Michael Newman/Photo Edit, 63, 65
Brandy Noon, 7, 45
© North Wind Picture Archives, 19, 20
PhotoDisc, 9, 35, 40, 46, 49, 67, 69, 70, 80
© Reuters/Jeff Christensen/Hulton/Archive, 52
© Ann Telnaes, 51
© Staton R. Winter/Liaison/Getty Images, 90

Library of Congress Cataloging-in-Publication Data

Rackley, Jenny
   Nicotine / by Jenny Rackley.
      p.   cm. — (Drug education library)
   Includes bibliographical references and index.
Summary: Discusses the history of tobacco, use, pop culture, health
and social problems, advertising, big business, liability, and continued
controversy.
      ISBN 1-59018-012-7 (hardback : alk. paper)
   1. Tobacco habit—Juvenile literature. 2. Smoking—Juvenile
literature. 3. Nicotine—Juvenile literature. I. Title. II. Series.
   HV5733 .R33 2002
   362.29'6—dc21

  2001005527

Contents

Foreword

The development of drugs and drug use in America is a cultural paradox. On the one hand, strong, potentially dangerous drugs provide people with relief from numerous physical and psychological ailments. Sedatives like Valium counter the effects of anxiety; steroids treat severe burns, anemia, and some forms of cancer; morphine provides quick pain relief. On the other hand, many drugs (sedatives, steroids, and morphine among them) are consistently misused or abused. Millions of Americans struggle each year with drug addictions that overpower their ability to think and act rationally. Researchers often link drug abuse to criminal activity, traffic accidents, domestic violence, and suicide.

These harmful effects seem obvious today. Newspaper articles, medical papers, and scientific studies have highlighted the myriad problems drugs and drug use can cause. Yet, there was a time when many of the drugs now known to be harmful were actually believed to be beneficial. Cocaine, for example, was once hailed as a great cure, used to treat everything from nausea and weakness to colds and asthma. Developed in Europe during the 1880s, cocaine spread quickly to the United States where manufacturers made it the primary ingredient in such everyday substances as cough medicines, lozenges, and tonics. Likewise, heroin, an opium derivative, became a popular painkiller during the late nineteenth century. Doctors and patients flocked to American drugstores to buy heroin, described as the optimal cure for even the worst coughs and chest pains.

As more people began using these drugs, though, doctors, legislators, and the public at large began to realize that they were more damaging than beneficial. After years of using heroin as a painkiller, for example, patients began asking their doctors for larger and stronger doses. Cocaine users reported dangerous side effects, including hallucinations and wild mood shifts. As a result, the U.S. government initiated more stringent regulation of many powerful and addictive drugs, and in some cases outlawed them entirely.

A drug's legal status is not always indicative of how dangerous it is, however. Some drugs known to have harmful effects can be purchased legally in the United States and elsewhere. Nicotine, a key ingredient in cigarettes, is known to be highly addictive. In an effort to meet their bodies' demands for nicotine, smokers expose themselves to lung cancer, emphysema, and other life-threatening conditions. Despite these risks, nicotine is legal almost everywhere.

Other drugs that cannot be purchased or sold legally are the subject of much debate regarding their effects on physical and mental health. Marijuana, sometimes described as a gateway drug that leads users to other drugs, cannot legally be used, grown, or sold in this country. However, some research suggests that marijuana is neither addictive nor a gateway drug and that it might actually benefit cancer and AIDS patients by reducing pain and encouraging failing appetites. Despite these findings and occasional legislative attempts to change the drug's status, marijuana remains illegal.

The Drug Education Library examines the paradox of drugs and drug use in America by focusing on some of the most commonly used and abused drugs or categories of drugs available today. By discussing objectively the many types of drugs, their intended purposes, their effects (both planned and unplanned), and the controversies surrounding them, the books in this series provide readers with an understanding of the complex role drugs and drug use play in American society. Informative sidebars, annotated bibliographies, and organizations to contact lists highlight the text and provide young readers with many opportunities for further discussion and research.

Introduction

Nicotine's Iron Grip

Nicotine is a chemical that is found naturally only in the leaves, seeds, and roots of the tobacco plant—*Nicotiana tabacum*. Tobacco, which is the dried leaf of the tobacco plant, has a unique distinction. It is one of the few substances legally available that can harm or even kill when used as the manufacturer intends. Nicotine is an extremely potent poison. It has been known to be a poison since the 1800s, when scientists first extracted nicotine from tobacco. Experiments in the mid-1800s showed that a few drops in the mouth would instantly kill a small animal like a cat. For years, nicotine was used as an agricultural insecticide.

In larger amounts, nicotine can poison humans. Too much nicotine can cause rapid pulse, vomiting, collapse, and even death. In the small amounts that are in tobacco, nicotine will not cause death instantly, but it still is a poison. It affects the body and the brain. Over time, either nicotine or other chemicals in tobacco or tobacco smoke can cause irreparable damage to a tobacco user's body.

There are a number of chemicals in tobacco, but nicotine is the chemical responsible for addiction. When people are addicted to nicotine, they feel as though they must have it on a regular basis in order to feel normal and functional. Other chemicals in nicotine can react with receptors in a tobacco user's brain to cause pleasurable feelings. Nicotine is estimated to be as addicting as heroin or cocaine.

Nicotine has been recently found to create a cancer-causing by-product. Many other chemicals in both tobacco smoke and smoke-less tobacco also cause harmful effects to the user. Cigarette smoke contains thousands of chemicals, about half of which are poisonous substances and over forty of which cause cancer. The Environmental Protection Agency (EPA) designates cigarette smoke as one of the most potent carcinogens. Scientists are still studying the effects of all of the different chemicals in cigarette smoke.

Most people who use nicotine are smokers. Tobacco can be smoked in a variety of ways—cigarettes, cigars, pipes, kreteks

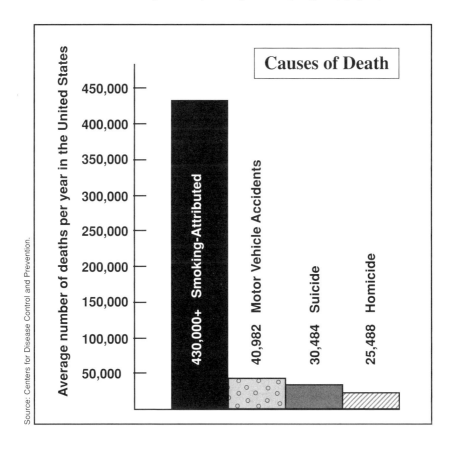

(clove cigarettes), or bidis (imported, candy-flavored cigarettes wrapped in leaves of a plant native to India). Smoking can cause chronic bronchitis, lung cancer, emphysema, heart and circulatory disorders, and other conditions. In fact, it is one of the nation's top killers—studies have shown that smoking causes over 430,000 deaths annually. It causes more deaths than alcohol, homicide, AIDS, car accidents, and illegal drugs combined.

Some people think that smokeless tobacco is safer. Snuff is a finely cut, powdered tobacco. Moist snuff can be put between a user's lower lip and gum or between the cheek and gum. Some forms of snuff are inhaled through a nostril. Chewing tobacco, also called chew, is formed of long strands of tobacco, and a wad of it is placed between the cheek and gum. Smokeless tobacco, which has many of the same carcinogenic chemicals as cigarette smoke, has been made popular in the past by sports figures. Many of those sports figures are now chewing bubble gum, because they have seen the negative effects of tobacco on their team-mates—a number of them have suffered from cancers of the mouth and oral cavity.

Tobacco companies have tried to develop "healthier" cigarettes, with lower nicotine and less harmful ingredients. However, most attempts to market healthier tobacco products have failed. "Light" cigarettes have a lower nicotine content, but smokers will inhale more deeply or more often to obtain the nicotine they need to feed their addiction.

Most adults who use nicotine began smoking as teenagers or children. Young people start using tobacco for many reasons, including stress, wanting to be more sophisticated or mature, or because their friends or parents smoke. Those who start smoking often continue because of nicotine's addictive nature. Many smokers truly enjoy their cigarettes. Says Jerry Thomas, a truck driver and a longtime smoker, "If I am going to die anyway, I might as well go out doing something I enjoy."[1]

Smokers find that the effects of cigarettes are pleasurable, and they feel "sharper" or more in control when they are using nicotine. Some people feel more comfortable in social situations when

Most adult smokers took up the habit in their teens, and sometimes younger.

they have a cigarette. Many smokers believe that smoking can help them feel more confident and less stressed. Still others believe that tobacco use will help them maintain a desirable weight.

Most of the long-term negative effects of using tobacco are difficult for people to recognize when they first start. Because it takes a long time for some of the tobacco-related illnesses to become apparent, most people initially ignore the health effects.

Becoming ill forty or fifty years down the road may not dissuade a child or teenager from starting to smoke. Many believe they will easily be able to stop smoking later, when they go to college or into the workplace. Yet 89 percent of adults who now smoke daily began using cigarettes before they were eighteen. In addition, 90 percent of smokers surveyed wished they could find a way to quit, even if they found it pleasurable to smoke.

For years, the tobacco industry denied the health risks of tobacco use. However, the tobacco industry has recently admitted it knew of the dangers all along. Bennett LeBow, owner of the Liggett Group (which produces Eve, Lark, and Chesterfield cigarettes) was the first to formally acknowledge that the tobacco industry has actively marketed to young people, that cigarette smoking is dangerous, and that nicotine is addictive. In a statement he made on March 20, 1997, he said,

> We at Liggett know and acknowledge that, as the Surgeon General and respected medical researchers have found, cigarette smoking causes health problems, including lung cancer, heart and vascular disease and emphysema. We at Liggett also know and acknowledge that as the Surgeon General, the Food and Drug Administration, and respected medical researchers have found, nicotine is addictive. . . .
>
> Liggett acknowledges that the tobacco industry markets to "youth," which means those under eighteen years of age. . . . Liggett condemns this practice.[2]

Because of the lawsuits against them, the tobacco companies have agreed to pay money to the states in compensation for expenses of sick tobacco users. In addition, they have agreed to run antismoking ads targeted toward young people and direct new ads toward adults only. Other lawsuits are still in progress.

Tobacco and the tobacco industry are being assaulted on many fronts. Many things have changed in recent years. There is still a turbulent legal climate; tobacco use in the United States has lost social acceptance; and there continues to be myriad new studies about tobacco's health effects.

In 1998, it was reported that almost 50 million Americans smoke. Although tobacco use has declined in the last few decades, millions of lives—tobacco users and their families, coworkers, and friends—are influenced by tobacco, and of course, nicotine. Nicotine's iron grip keeps a strong hold on those who use tobacco, in any of its forms.

Chapter 1

Tobacco's Roots in the Western World

Nicotine is a naturally occurring, addictive chemical found in the leaves of the tobacco plant. Although tobacco was unknown to Europe before Columbus's historic journey, it grew naturally in the Americas and was used by many native tribes. In the five centuries after Columbus brought tobacco back to Europe, tobacco use spread to every known culture.

Governments have reacted in many different ways to the proliferation of tobacco in their countries—from taxation, penalties, or even death to encouraging tobacco's production so that the government could profit from it. With automation and advertising, tobacco use grew exponentially throughout the first half of the twentieth century. Almost 50 million Americans inhale or ingest tobacco in the form of cigarettes, smokeless tobacco, or other tobacco products. It is one of the few legal drugs available today, and its popularity exceeds that of other drugs.

A Gift from the New World

In the fifteenth century, many Europeans had a desire for expansion and trade, including Italian-born explorer Christopher Columbus. After settling in Spain, Columbus joined in the search

leaders and healers called shamans used tobacco to treat countless diseases and disorders. Tobacco leaves were used as poultices over wounds or snakebites. The smoke was used to ease the pain of childbirth or injury. Some shamans used the smoke to drive out evil spirits that were believed to cause illnesses.

Often tobacco was smoked with great ceremony. Groups of elders or important tribal members gathered, sometimes in connection with a religious ritual, and shared a common pipe. The elders or shamans believed that the mind-altering effects of tobacco provided a connection with the gods. Many believed that tobacco had miraculous powers, and by using tobacco, they could ensure a variety of good fortunes, from sufficient rainfall to plentiful harvests.

Tobacco Spreads in the Old World

After Columbus brought tobacco back to Europe, tobacco usage grew slowly and steadily. Some people tried it because a friend or neighbor recommended smoking. Others believed that it had health benefits and restorative properties. Many believed that smoking cured all manner of illnesses, from gout to asthma.

Sir Walter Raleigh was an avid proponent of smoking. Here a servant, believing Sir Walter to be on fire, prepares to douse him with water.

Tobacco in the Roanoke Colony

Thomas Hariot was part of the colonization attempt initiated by Sir Walter Raleigh on Roanoke Island. In his 1588 book titled *A Briefe and True Report of the New Found Land of Virginia,* Hariot describes how the natives used tobacco and how the Europeans who were convinced of its many health benefits adopted it:

> There is an herb called uppowoc, which sows itself. In the West Indies, it has several names, according to the different places where it grows and is used, but the Spaniards generally call it tobacco. Its leaves are dried, made into powder, and then smoked by being sucked through clay pipes into the stomach and head. The fumes purge superfluous phlegm and gross humors from the body by opening all the pores and passages. Thus, its use not only preserves the body, but if there are any obstructions, it breaks them up. By this means, the natives keep in excellent health, without many of the grievous diseases which often afflict us in England.

> This uppowoc is so highly valued by them that they think their gods are delighted with it. They make holy fires and cast the powder into them as a sacrifice. If there is a storm on the waters, they throw it up into the air and into the water to pacify their gods. Also, when they set up a new weir for fish, they pour uppowoc into it. And if they escape from danger, they also throw the powder up into the air. This is always done with strange gestures and stamping, sometimes dancing, clapping of hands, holding hands up, and staring up into the heavens. During this performance they chatter strange words and utter meaningless noises.

> While we were there we used to suck in the smoke as they did, and now that we are back in England we still do so. We have found many rare and wonderful proofs of the uppowoc's virtues, which would themselves require a volume to relate. There is sufficient evidence in the fact that it is used by so many men and women of great calling, as well as by some learned physicians.

Sir Walter Raleigh, a sixteenth-century soldier, businessman, explorer, and writer, did much to popularize smoking among the upper classes in Europe. Raleigh began smoking after trying tobacco that had come to England from a seized Spanish ship. Raleigh soon became an avid proponent of tobacco. And, because he was highly influential and distinguished, many other English noblemen gave smoking a try. Queen Elizabeth I herself is reported to have tried smoking at Raleigh's behest. Yet smoking was not very common—a servant once thought Raleigh was on

fire when he saw Raleigh surrounded by smoke, and he poured a
bucket of water over Raleigh's head.

Another European who had great influence over tobacco use
was Jean Nicot, the French ambassador to Portugal in the mid-
sixteenth century. The active ingredient in tobacco, nicotine, was
named after this very outspoken supporter of tobacco. Nicot ob-
served how the Portuguese valued tobacco for its healing powers.
He wrote to the French nobility telling them of the great won-
ders of tobacco. He even sent seeds to the French courtiers to try.
The queen of France tried tobacco for her migraine headaches. It
did not take long for the French to become believers in the power
of tobacco. Within forty years, tobacco use had spread to most
French citizens, who believed that its disinfectant and antiseptic
powers could cure or prevent all manner of ailments. Some even
believed tobacco was effective against the plague.

By the early seventeenth century, the Europeans had introduced
tobacco to many other nations and territories. Sailors wanting a
regular supply of tobacco left seeds along their trade routes. Local
farmers began to cultivate the tobacco and develop local crops.
Later, when the sailors returned to their ports of call, they had tobacco waiting for them. Europeans, finding tobacco a valuable

*Jean Nicot, the French
ambassador to Portugal
in the mid-1500s, lent his
surname to the active
ingredient in tobacco:
nicotine.*

new commodity, began to offer it in trade. Tobacco use spread throughout Africa, Asia, and the Middle East.

Serious Consequences

Although tobacco was quite popular, and believed by many to be beneficial, it generated controversy and conflict almost from the beginning. In 1621, Robert Burton wrote in *The Anatomy of Melancholy* about tobacco's values and vices:

> Tobacco, divine, rare, superexcellent tobacco, which goes far beyond all the panaceas, potable gold, and the philosophers' stones, a sovereign remedy to all diseases . . . but as it is commonly abused by most men, which take it as tinkers do ale, 'tis a plague, a mischief, a violent purger of goods, lands, health, hellish, devilish and damned tobacco, the ruin and overthrow of body and soul.[3]

Many monarchs and governments tried to stem the flow of tobacco into their nations because they recognized that the health benefits that had been promoted were in fact wrong. In 1665 an experiment was done in which a cat was fed distilled oil of tobacco. The cat died within seconds. Others noted increased respiratory ailments and diseases among tobacco users. However, despite this evidence, many people still believed that tobacco use was a healthy practice.

In addition, many monarchs and other leaders did not like how their citizens became enslaved to their smoking habit. Those who used tobacco were addicted to it and used it despite threats of serious penalties. The sultan of Turkey blamed a serious fire on careless smokers. Religious leaders called tobacco offensive to both God and man. One smoker was jailed after being accused by a priest of having evil spirits coming out of his mouth.

Laws were passed to deter smoking and limit tobacco import. Some leaders placed exorbitant taxes on tobacco. In 1604 King James I of England published an antismoking tract called *A Counter-Blaste to Tobacco,* in which he condemned the practice of smoking as "a custom loathsome to the eye, hateful to the nose, harmful to the brain, dangerous to the lungs, and in the black stinking fume thereof, nearest resembling the horrible Stygian smoke of the pit that is bottomless."[4]

The leaders of some nations had more severe penalties for smokers. The Hindustani emperor ordered smokers' lips to be split as a punishment and deterrent. Those who sold tobacco in China were beheaded. Russian smokers were banished to Siberia. Smokers in Turkey met the wrath of the sultan, who had been dismayed by the fire; those caught smoking had their pipes driven through their noses just before they were beheaded.

Governments Go for the Gold

Like many illicit substances, though, nothing diminished the growing popularity of tobacco. Just as Columbus realized he had little power over his crew's tobacco use, most leaders recognized that they could not stem the flow of tobacco products into their nations. Many instead decided to profit from tobacco by increasing taxes on the sale and import of tobacco products.

Tobacco grown in the Spanish colonies of Cuba, Mexico, and the West Indies was quite popular in the sixteenth and early seventeenth centuries. The Europeans favored the taste and the Spanish tobacco soon cornered the market. To compete with the Spanish, the English thought it was to their benefit to have their own colonies in America begin producing superior tobacco.

John Rolfe, a leader of the Jamestown settlement in Virginia and the man who later married the daughter of the Powhatan tribe's chief, Pocahontas, obtained some tobacco seeds from a Spanish colony. He planted those seeds in Virginia, and soon the colony was producing some of the finest-quality and most flavorful tobacco available.

To encourage its Virginia colony to become a competitive force in the tobacco-producing industry, England offered land grants and low-priced ocean passage to those wanting to establish tobacco operations. Farmers in England were prohibited from growing tobacco so that a stronger market would develop for the Virginia crops. Penalties and duties were imposed on those importing tobacco into England from anywhere other than Virginia.

Cultivating tobacco in colonial Virginia. The colony was noted for the high quality and excellent flavor of its tobacco.

In exchange for the guaranteed market for Virginia tobacco, growers agreed to sell exclusively to British merchants. The tobacco could not be shipped on any ship not flying a British flag. If merchants in other countries wanted Virginia tobacco, the British merchants resold it to them.

Virginia produced more and more tobacco, and with England as its only customer it became greatly dependent on this trade relationship. Farmers in Virginia grew few other crops—just what was needed to provide food for themselves and their workers. British merchants brought supplies to the farmers, but often charged what the growers considered unfair prices. Many producers had borrowed money from British firms to establish their tobacco operations, buy supplies, and secure slave labor to work on their plantations. Although tobacco production was quite successful, Virginia farmers were heavily indebted to England. With an abundant supply of tobacco, the prices that the British would pay dropped. Consequently, with lower incomes and high debt loads, the profits of the colonial tobacco farmers all but disappeared.

However, this situation changed drastically during the Revolutionary War. Tobacco was not the only item that was restricted in the colonies; England controlled many other markets as well. For example, England prohibited colonists from making metal tools so there were guaranteed markets for high-priced, British-made tools. Other goods were highly taxed. The colonists asked the English for relief, but the English responded by imposing more taxes. The angry colonies declared their independence from England and fought the Revolutionary War to secure that independence. Virginia and other tobacco-producing colonies hoped that the Revolution would provide freedom from all the restrictive trade practices of the English king and relief from their crippling debt.

The war did as they hoped. In fledgling America, tobacco growers could begin anew. No longer burdened by their debts, the growers looked for new markets. However, international tobacco trade had slowed down after the war. As a result, many tobacco producers started providing tobacco for domestic tastes—

Tobacco is loaded onto a ship in Virginia. The ship is bound for England, which exercised strict control over colonial tobacco farmers.

they grew the types of tobacco and supplied the tobacco products that their new compatriots favored.

Chew, Chew!

Many early colonists smoked pipes or used snuff, a powdered or finely cut form of tobacco. While most Europeans preferred to inhale powdered snuff, the colonists preferred moist snuff, which would dissolve slowly when placed between the lower lip and gum. After the Revolutionary War, however, American tobacco preferences turned toward chewing tobacco. Chewing tobacco, also called chew or chaw, was already popular with sailors and others who worked outdoors. It was sweetened, convenient to carry, and kept the hands free so that one could chew tobacco and work at the same time. Sailors and others did not have to worry about wind and weather causing interference with a lighted cigar or pipe.

Although chew was popular, its use had some unpleasant consequences. When people chew tobacco, they must spit out the tobacco juice. Spittoons, containers to catch brown-colored spit and used chewing tobacco, were placed everywhere. If chewers could aim well, they hit the spittoon. Otherwise, when they missed, carpets and other items became soiled. In his book *Ashes to Ashes* Richard Kluger said that there was "a veritable stream of tobacco juice [which] filled the air throughout much of the nineteenth century, targeted at the [spittoon] but at least as frequently darkening carpets, walls, draperies, and trousers."[5] Outside, people could and would spit anywhere. Because of this habit, visitors from other countries thought the Americans were messy and filthy.

Use of chew dramatically lessened in the mid-nineteenth century when it was discovered that tuberculosis could be transmitted through spit. In some places it even became illegal to spit in public. As smokeless tobacco became less popular, use of other forms of tobacco increased.

From Cigars to Cigarettes

Native Americans had smoked mostly cigars, which were made by rolling dried tobacco into larger tobacco leaves. Cigars became quite popular with many American smokers in the mid-nineteenth century. In Europe, another form of tobacco was in use. Spanish smokers were trying *cigaritos*, or "little cigars." To make *cigaritos*, shredded tobacco was placed on tan paper and was then rolled into a thin tube. The ends were twisted to keep the tobacco in. The *cigaritos* were popular but expensive and time-consuming to make. Many people simply bought the ingredients and rolled their own.

In France and England, these cigarettes (French for "small cigars") were not yet popular except with the poor. They could sift through garbage to find cigar stubs and other tobacco products and reroll them with their own paper into cigarettes. Cigarettes actually had a bad reputation in much of Europe at first. They were said to be dirty and contaminated. Some people thought the cigarettes might be tainted with dung, urine, or spit.

However, the Crimean War (1853–1856) changed Europeans' perceptions about cigarettes. Because they were lightweight, easy-to-use, portable, and inexpensive, cigarettes became popular among European soldiers. After the war, London tobacco producer Philip Morris began supplying cigarettes to satisfy the new tastes of the soldiers.

Tobacco in the Trenches

As nicotine use was becoming even more widespread in all its different forms, eventually one tobacco product cornered the market. Cigarettes, the most popular form of tobacco, were made even more popular during times of conflict. Wars helped shape the choices that tobacco users ultimately made. Wars were very difficult times for soldiers, who were often lonely, uncertain of what to expect, and far from home. Using tobacco gave soldiers something to do and helped them cope with their anxieties.

In the American Civil War, cigarettes became extremely popular with soldiers on both sides of the conflict. They were easy to smoke and fit in with the lifestyle of the soldier, who had only limited time for smoking and little money for pipes or cigars. Later, other wars popularized tobacco and cigarettes even more. In both World War I and World War II, cigarettes were included in daily

A wounded World War II soldier (center) smokes a cigarette. During both world wars, cigarettes were included in soldiers' rations.

rations. They were considered at least as important as food. In fact, according to writer Richard Kluger, during World War II President Roosevelt declared tobacco "an essential wartime material"[6] and allowed tobacco growers exemptions from going to war.

All of these military conflicts did much to spread tobacco use. Besides being given to soldiers by their governments, cigarettes were available at low cost to the armed forces. Additionally, charitable organizations like the Red Cross and the YMCA delivered cigarettes free-of-charge to the military. Tens of thousands of young soldiers went to war as nonsmokers and returned home with an addiction they would have for a lifetime. Modern wars, and the government support of tobacco products during those wars, did more to popularize cigarettes and tobacco use than did any other single event in history. By 1949, over half of all American men smoked tobacco.

Hand-Rolled Versus Machine-Made

Although cigarettes increased in popularity over the years, mainly as a result of war, they were hard to make commercially. In the late 1800s most smokers rolled their own cigarettes. Some companies produced ready-made cigarettes, which were rolled by hand in factories, usually employing large numbers of young women. However, these did not have a very strong market since they were expensive to produce.

In 1880 one cigarette company sponsored a contest with a prize of the then-huge sum of seventy-five thousand dollars for the first person to develop a machine to roll cigarettes automatically. Mechanical production had been tried before, but it had not worked very well. The paper tore or wrapped the cigarettes unevenly, and the shredded tobacco got into everything and caused the equipment to break down.

James Albert Bonsack, an inventor and son of a plantation owner from Lynchburg, Virginia, had been tinkering with mechanical production of cigarettes since he was a teenager. In 1880, when he was twenty-one, Bonsack obtained a patent for his automatic cigarette-rolling machine. The company offering the prize rejected the machine, even though it could produce over seventy thousand cigarettes in a single shift. However, it did not operate as efficiently as Bonsack had hoped.

Early-twentieth-century female workers roll cigarettes by hand.

Although the cigarette company ultimately rejected Bonsack's machine, he did not give up. Within two years, he sold five of his machines on a trial basis to another company, run by James Buchanan "Buck" Duke. Duke's engineers helped Bonsack solve the efficiency problems of his machine on the condition that Duke would get royalties on any machines sold to competitors.

The Bonsack machine was a gold mine for Duke. The new machine-rolled cigarettes could be produced for less than half the cost of hand-rolled cigarettes. Operating even more efficiently, Bonsack's machine was now able to produce more cigarettes than forty workers could produce by hand in the same time period. In 1884, Duke produced 744 million cigarettes—more than the national total from all manufacturers in the previous year.

Young People and Women: Advertising Targets

Duke's cigarettes were rolling out of his factory so quickly that he needed to find ways to make people want to buy them. A competitor was already inserting small trading cards in their cigarette packs. Buck Duke decided he needed to do more to attract customers. He began aggressive marketing campaigns that set the stage for cigarette advertising and the growth of U.S. tobacco companies for the next hundred years. By 1899, he was spending over eight hundred thousand dollars annually on marketing.

Tobacco companies started marketing to young people almost from the start. Cigarette trading cards often featured sports figures or actresses. These items were very popular with young people, who would either purchase the cigarettes themselves for the promotional item or convince their parents to purchase them. The demand by minors grew steadily, and retailers were willing to meet that demand.

Tobacco companies soon realized they were ignoring another whole market: the female smoker. In the United States in the early part of the twentieth century, women rarely smoked. It was illegal in some locations, and considered improper in most. In some circumstances, a woman could smoke privately at home, but never

By the 1920s, cigarette advertisers began targeting women, implying that smoking was an effective means of weight control.

out in the open. Women were expelled from schools and even arrested for smoking in public.

The promotional items inside cigarette packs were a good way for the tobacco companies to focus on women. In 1912 they started offering silk rectangles in each pack. These rectangles could be sewn together to make a scarf or quilt cover.

By the 1920s, cigarette ads started to be targeted toward women. Ads began to show women enjoying life by enjoying cigarettes. Some brands were specifically designed for women, and the tobacco industry played on social fears to encourage women to smoke. They started billing cigarettes as an effective way to stay slim. In 1928 the American Tobacco Company, which made Lucky cigarettes, promoted them with the slogan, "To stay slender, reach for a Lucky. A most effective way of maintaining a trim figure."[7] Another Lucky ad read, "Reach for a Lucky instead of a sweet."[8] The ads encouraged women to pick up a cigarette for weight control, and more and more women did.

Women usually smoked indoors, but tobacco companies wanted to make it more acceptable for them to smoke outdoors. In 1929 the American Tobacco Company sponsored a march of New York debutantes and fashion models. These women "declared their independence" from the social stigma of women smoking by dressing as the Statue of Liberty and holding lighted cigarettes up high like "torches of freedom."

Tobacco advertisements also focused on the independent woman during World War II. Women as well as men became widely addicted to cigarettes. Many women began to work outside the home to fill jobs left vacant by men fighting overseas. Cigarette advertisements targeted these independent women, working to help the war effort and support their families. Ads featured patriotic women smoking in the workplace, riveting panels on airplanes, or doing other wartime jobs. In 1949, about 33 percent of all American women smoked tobacco.

Tobacco's Turning Point

After the world wars, tobacco use continued to build in popularity. In many circles, it seemed almost more common to smoke than not to smoke. In 1963, American adults were consuming 4,345 cigarettes per capita. Both men and women increased their cigarette consumption until the landmark 1964 report of the surgeon general titled *Smoking and Health.*

President John Kennedy commissioned this report to determine whether there was any truth to the report of negative health effects of smoking. The resulting report was a culmination of many months of study and research, and it stunned the nation. Contrary to what the tobacco companies had said, it reported some very serious hazards of tobacco use. The surgeon general's report changed the way people thought about tobacco.

Over the last five decades, researchers have published study after study detailing how tobacco affects people's health. As a result, tobacco consumption dropped significantly. When tobacco users found out what their habit had been doing to their health, it caused many to quit immediately. Others were still hooked by nicotine and continued to use tobacco despite knowing there were health risks. Although overall the smoking rate has declined, in 1999, American adults still consumed 2,146 cigarettes per capita. The nicotine in tobacco keeps people addicted and their addiction keeps them using tobacco.

Chapter 2

Addicted: In the Lungs and in the Brain

Nicotine is an addictive substance. When nicotine enters the bloodstream, it alters both the mind and the body. Nicotine, along with the other chemicals in tobacco, produces some pleasurable effects—but there are also very serious long-term health implications to those using tobacco products and those who live or work with tobacco users.

Today adults continue to use cigarettes much more often than other forms of tobacco. According to a 1998 survey, 24 percent of American adults smoke cigarettes (26 percent of men and 22 percent of women). The figure is lowest among the most educated adults, with only 11 percent of college-educated adults smoking. It is highest (37 percent) among those who were not high-school graduates. Some types of tobacco are used almost exclusively by men—4 percent of men use cigars, 2 percent use pipes, 3 percent use chewing tobacco, and 3 percent use snuff. Less than half of 1 percent of women use tobacco in these ways.

Although tobacco users and medical professionals have known for generations that nicotine was addictive, it is only recently that studies have proved (and tobacco companies have admitted) this fact. As an addictive substance, tobacco affects people's behavior, causing them to make choices that are harmful to their health and the health of others.

In the Bloodstream

When any form of tobacco is used, nicotine is absorbed into the bloodstream. When a person places moist snuff or chew in between the cheek and gum or inhales snuff through the nose, nicotine immediately moves into the small blood vessels in the mucous membranes of the mouth and nose. These mucous membranes in the oral and nasal cavity are thin and allow nicotine and other chemicals to be quickly absorbed into the bloodstream.

Although some people use smokeless tobacco products like chew or snuff, most tobacco users inhale the tobacco smoke. As humans inhale, oxygen is drawn into the lungs. The bronchial tubes, or airways inside the lungs, look like large upside-down stalks of broccoli. The larger parts of the stalk direct the air down

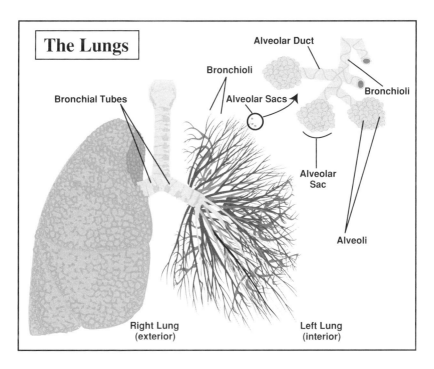

into the smallest sections. Alveoli are the small sacs found at the end of these smallest branches. They cover a huge area, with more than ninety times more surface area than the skin. It is in these alveoli that fresh oxygen is supplied to the blood during inhalation and carbon dioxide is taken out of the blood during exhalation.

When a smoker takes a puff on a cigarette, a mixture of oxygen, nicotine, and carbon monoxide, among other chemicals, are brought into the arterial bloodstream through the lungs. The arterial bloodstream supplies fresh, oxygenated blood to the brain and all the body's tissues. The brain and body need oxygen to survive and function normally. But nicotine and other chemicals from tobacco replace some of the oxygen sent into the bloodstream and, once in the bloodstream, the nicotine moves very quickly to the brain. In fact, nicotine arrives in the brain eight to ten seconds after it is inhaled or ingested.

In the Brain

When the chemical-laden arterial blood flow reaches the brain, nicotine begins promoting the production of dopamine. Dopamine is a neurotransmitter, or a chemical in the brain that carries messages between nerve cells, and is part of the complex neural circuitry that helps the body perceive pleasure. Dopamine is normally released by the brain when doing something necessary for survival (like eating) or during other pleasurable events (such as giving someone a hug). So, when nicotine enters the brain and triggers dopamine production, the tobacco user perceives that his or her action was pleasurable.

Nicotine triggers dopamine production by imitating another neurotransmitter, acetylcholine. When certain receptors on dopamine-producing nerve cells are filled with acetylcholine, dopamine is produced in moderate and controlled quantities. Nicotine is similar enough to acetylcholine that it can lock into its receptors. By doing so, it fools the body into producing elevated levels of dopamine. Alcohol, heroin, and cocaine also promote dopamine production in the brain.

Nicotine also reacts in different parts of the brain to produce other chemicals. Glutamate, the neurotransmitter that enhances memory and learning, is released when nicotine is used. One effect of this enhanced memory is that the brain remembers the pleasure it experienced as a result of the dopamine and wants to experience that pleasure again.

Endorphins, which are known as the body's natural painkiller, are also released in the presence of nicotine. Runners experience a "runner's high" from the release of endorphins. This gives runners a mental advantage—they do not have to worry about small aches and pains while they complete the race. The effects of endorphins might similarly contribute to the mental edge tobacco users feel after taking nicotine into the body.

Nicotine has several other complex physiological effects. It affects central nervous system functions, thought processes, and behavior. Some people report that nicotine separates them from their emotions, and with nicotine they can think more logically and work better. This effect helps explain why some people feel better able to cope when they use tobacco. As David Sanford, a longtime smoker, puts it, "I work better when I can smoke. . . . Smoking definitely helps me concentrate, and it has many anti-depressant effects (it helps me cope with my multiple sclerosis)."[9]

Other chemicals released by nicotine's actions in the brain help the tobacco user feel alert and energized. But for all the functions that are temporarily improved, there is a very real danger of current and future health problems. The prolific writer Stephen King said, "I used to be faster than I am now; one of my books . . . was written in a single week. . . . I think it was quitting smoking that slowed me down; nicotine is a great synapse enhancer. The problem, of course, is that it's killing you at the same time it is helping you compose."[10]

An Addictive Substance

The perceived benefits of tobacco are one of the reasons that people become addicted to using it. It changes the way people think and process information and how they cope with the difficulties around them. However, addiction is a complex process that

involves the chemical properties of the drug, its cost and availabil-
ity, the user's specific personality traits and genetic factors, the
user's peer and family influences, prevailing social attitudes, and
more. All of these factors mix together to help determine whether
a person who tries a drug will subsequently become addicted.

When someone is addicted to a drug, the use of that drug on a
regular basis becomes compulsive. This leads to a feeling that the
user of the drug cannot control the compulsion. Without the
drug, users start to experience a variety of uncomfortable symptoms, including powerful cravings. To alleviate these symptoms,
they continue to use the drug despite knowing about its health
implications. Steven Hoadley, a longtime smoker, started smoking
when he was twenty and later became addicted to alcohol, cocaine, and heroin. Eventually, because of his addictions, he found
himself homeless. He says,

> Tobacco was always there. . . . I stole my cigarettes from whatever source I
> could. In those days, smoke products weren't as guarded as they are today.
> Supermarkets and liquor stores . . . often kept a display rack on the
> counter. I would ask the clerk to get me a pint of whiskey and when he
> turned his back, I pilfered a pack or two.

If I was walking down a street and a car door was open and an open pack sat on the dash, they were mine. Basically, I got them however I could. They were never a priority at that time, as the other substances took precedence. But, nevertheless, I went to great lengths.

Addiction is a full time job.[11]

The kick that smokers experience from smoking increases each time certain portions of the brain are exposed to nicotine. Nicotine activates cells associated with memory and learning in the hippocampus, a part of the brain. Seeking this remembered "nicotine high," an addicted smoker continues to breathe poisonous and cancer-causing substances into his or her body.

Research has shown that it only takes one cigarette for the brain to recall the nicotine high. However, it still takes an effort to begin smoking cigarettes. More than other addictive drugs, nicotine can cause dizziness, nausea, and severe discomfort the first time it is used. The beginning smoker will usually cough and gasp for air as he or she is learning to smoke. A new smoker must have good motivation for continuing to try cigarettes.

Sometimes that motivation is external—beginning smokers may want to fit in with their friends, they may have heard smoking is good for stress or diet, or they may have a family member who smokes. They may also have been brainwashed by advertising into thinking that smoking is cool, sophisticated, or macho. Dr. Elizabeth Whelan, part of the American Council on Science and Health, stated that, "cigarette smoking has not only the potential to addict physiologically by means of nicotine after just a few smokes, but also the ability to compound that addiction through behavioral and psychological facets associated with the act of smoking a cigarette."[12] These behavioral or psychological factors, the habitual factors that drive a smoker to continue smoking, can be very hard for smokers to overcome.

Sometimes the motivation to smoke is internal. After a smoker experiences the effects of nicotine, he or she may want to continue to experience that feeling despite the difficult side effects. Additionally, even after just a few cigarettes, the new smoker who stops suddenly may feel withdrawal symptoms such as irritability or difficulty in concentrating or thinking. These symptoms, along with a craving for more nicotine, make it very hard to stop smoking. Christina Johnson, a longtime smoker, tells of her experiences when she ran out of cigarettes while visiting a friend. She says, "I was desperate! My body ached for it. . . . My friend . . . went around gathering cigarette butts and squeezed the last little bit of tobacco from . . . them . . . until he had enough to roll me a makeshift cigarette. It was probably the worst cigarette I've ever had but, at the time, you couldn't have told me that. I sucked it down in minutes."[13]

Smokers, however, often believe they are *not* addicted. Many smokers say they like to smoke but can quit anytime. And for a very few, that may be true. Some smokers seem to be able to go for days without a cigarette, smoke a few, and then resume their smoke-free status. This is very rare though. For most people, once their brain has felt that nicotine high, they have a desire and a real need to keep on using nicotine at any cost.

Danger, Danger!

Because of nicotine's addictive qualities, tobacco users take the drug again and again. The first dose of nicotine causes the user to feel instantly awake and alert; following that, there is a calm, relaxed feeling. However, the resting heart rate actually increases by two to three beats per minute after each dose of nicotine. Nicotine constricts the blood vessels, reducing the skin temperature and decreasing blood flow to the legs and feet. Thus the smaller blood vessels cause blood pressure to be increased, leading to lightheadedness and a rapid heartbeat. While these immediate effects are not harmful in themselves, repeated use causes damage to the body that takes place over the course of the tobacco user's lifetime.

Nicotine is a highly potent poison that has been used as an agricultural insecticide. Nicotine is so dangerous that 41 percent of seasonal farmworkers who harvest tobacco contract a severe case of what is known as green tobacco sickness. This occurs when tobacco field workers handle the moist tobacco leaves and absorb nicotine through the skin, causing symptoms such as nausea, vomiting, dizziness, headaches, and cramps.

Harmless-looking tobacco leaves contain deadly toxins when dried and smoked as cigarettes.

In addition to nicotine, many ingredients in tobacco itself are also toxic. For example, there are at least four thousand chemicals in cigarette smoke, two thousand of which are poisonous and over forty of which are verified to be carcinogenic. Besides tar, a sticky, brown substance deposited in smokers' lungs, ingredients of cigarettes include: acetone (found in nail polish remover); ammonia (a poisonous gas); arsenic (a rat poison); carbon monoxide (a poisonous gas found in car exhaust); hydrogen cyanide (a toxic gas used as a pesticide); DDT (a banned insecticide which caused serious harm to animals and the environment when it was in use); and formaldehyde (embalming fluid).

Tobacco producers add these chemicals in order to maintain consistent levels of nicotine in the product. Manufacturers' goals in manipulating nicotine levels are to cause people to become addicted to tobacco products.

The nicotine in tobacco does not have much to do with how much nicotine was in the leaves on the original plant. In fact, on the same tobacco plant, concentrations of chemicals in the different

Exercise and Smoking

Some people think those smokers who exercise rigorously are working to improve their physical condition and will therefore reduce their risk of heart attack. Yet a recent study showed that smokers who exercise have a five times greater risk of heart attack than nonsmokers do.

Normally when people exercise, their bodies release epinephrine, which speeds up their heart rate. Nicotine mimics epinephrine, so a smoker's heart is constantly signaled to speed up. The extra signals that nicotine causes are essentially false alarms, and the body stops reacting to these extra signals because it is being flooded with them. So, during exercise, when the body really does need a faster heart rate, a smoker's body may not recognize it.

A person with a diminished heart response rate may not get enough of the blood needed during a vigorous workout, which could lead to a heart attack. About one-third of heavy smokers showed a diminished response rate during an exercise test. Of men with an impaired heart response rate, almost one out of five had a heart attack and one out of ten died during the eight-year study.

In addition to increased risk of heart attacks, the lung capacity of smokers is reduced. Smokers also have reduced endurance and are often short of breath. "I can't smoke more than two cigarettes a day when I run daily," says nurse-midwife Eleanor Swift in an interview with the author. "Running is not compatible with smoking." Madeleine Armstrong found that her breathing dramatically improved about six months after she quit smoking. In an interview with the author she said, "I found . . . that my breathing when I play sports was so much better. Before, I did not see the difference because I had been smoking for so long that I did not know I had a problem."

leaves can vary widely, with the leaves toward the top of the plant having more nicotine and a higher concentration of other chemicals.

However, tobacco companies require a standard level of nicotine that is difficult to regulate in the natural product. So they use these chemical additives and other means to maintain the nicotine at optimal levels. Additives like ammonia increase the amount of nicotine available to the body. Other techniques include blending different tobacco leaves together, changing the cigarette or paper design, and changing the filter. The tobacco companies' goal in making a product with a standard amount of nicotine is to get the smoker hooked. Once hooked, they will likely be customers for a long time.

Although there are also some food and fragrance components in tobacco, no one has studied how all these added substances change chemically as they are burned or how these new chemicals affect the body. A natural chemical that may be safe as a food product can be carcinogenic in its combusted state.

Tobacco's Toll

All tobacco users send harmful chemicals through their body. Because there is not enough time for the body to repair itself between assaults, using tobacco is like contributing to death or disease in daily doses.

Manda Djinn, a singer and writer who lives in Paris, France, began her three-pack-a-day habit at the age of eleven and sang for years in smoky nightclubs. She says, "I am left with asthma and

Kelley. © Steve Kelley. Reprinted with permission.

Smokeless, not Harmless

Bill Tuttle, an outfielder for the Tigers, A's, and Twins baseball teams, chewed tobacco for almost forty years. In 1993, Tuttle's wife noticed a very large lump in his mouth. He went to the doctor and learned that the lump was malignant—Tuttle had cancer.

Over the course of six surgeries, Tuttle lost most of his face. The surgeons kept removing more and more bones and other tissue because the cancer continued to spread. Tuttle's jaw was eventually cut away, and the plastic surgeons had to work hard to reconstruct his face. They took part of his skull and flipped it down to create a new cheekbone. Huge pieces of skin from his chest and other areas of his body were used to cover the places where his cancerous facial skin had to be removed. He had lost his teeth earlier because of dental problems related to his chewing habit. He lost almost seventy-five pounds. Nerves had been cut in some of his many operations so he had no arm strength and could not even open a bottle.

After his cancer took so much from him, he decided to give something back to other ballplayers. Before his death in 1998, Tuttle gave talks around the country to ballplayers and the public. First the audience was shown a picture of Tuttle as a ballplayer in his prime. Then Tuttle would

appear and the audience could see firsthand the effects of smokeless tobacco. Some of the ballplayers who were chewing at the start of his talk would shamefacedly stop by the end. Because he was not afraid to speak out, Tuttle helped other ballplayers and smokeless tobacco users, from Little Leaguers to major leaguers, become tobacco free.

Baseball player Bill Tuttle's health was ravaged by his use of chewing tobacco.

respiratory allergies. . . . I'm still singing . . . but smoke anywhere in my vicinity makes me short of breath, stings my eyes and irritates my nose. . . . We did not have the facts [years ago] about nicotine that we have today."[14]

Countless scientific studies have shown how nicotine and tobacco smoke affect human health. Nicotine and the other chemicals in tobacco damage many body systems, including the immune system. Tobacco users increase their risk of cancer, heart disease, circulatory disorders, ulcers, and stroke. Nicotine mimics epinephrine, which means a smoker's body is constantly signaled to speed up its heart. So the heart has to work harder than it should and may not respond well when an increased heart rate is needed by the body.

A smoker's lungs are damaged and lose their elasticity because of the many chemicals in cigarette smoke: tar coats the delicate lung surfaces and smoke damages the cilia (protective hairs) in the respiratory tract. Because of this, smokers have more frequent occurrences of colds and respiratory infections, bronchitis, emphysema, and asthma. Other effects that have been linked to tobacco use are rheumatoid arthritis and hearing or vision loss. Tobacco interferes with the absorption of calcium and other nutrients, so over time tobacco users have lower bone density and an increased risk of osteoporosis. Because nicotine reacts with brain chemicals, it alters the way a person thinks, in some cases causing depression, impaired thinking, and dementia.

It is estimated that smoking costs an average of eleven minutes of life for each cigarette—a pack of cigarettes will shorten a smoker's life by three hours and forty minutes. Clive Bates, director of Action on Smoking and Health (ASH) said, "As if that's not bad enough, smokers are likely to die a more painful death and spend longer being ill while they are alive."[15]

Even though scientists knew nicotine was toxic and addictive, it was once believed nicotine itself did not cause cancer. Recent studies have proved that wrong—nicotine does react in the body to create a cancer-causing by-product. Scientists have recently added nicotine itself to the list of known cancer-causing chemicals in tobacco.

Even smokeless tobacco has its problems. One study showed that 46 percent of baseball players who used snuff or chew had oral lesions, as compared to 1.4 percent of nonusers. Most lesions were not cancerous, but the ballplayers in this study were young and the median time they had used tobacco was only five years. It is not known how many of those lesions have changed and become cancerous in the years following the study.

Tobacco is one of the United States's top killers—it is estimated that tobacco use causes over 430,000 deaths annually. Broken down by day, more than 1,100 people die each day of smoking-related causes—enough to fill three jumbo jets. Tobacco use directly causes more deaths than alcohol, homicide, AIDS, car accidents, and illegal drugs combined.

Susannah Tate quit smoking almost two decades ago, but her family has been greatly affected by tobacco use. She tells of tobacco's toll: "My older sister's lungs are so damaged she is on oxygen and has had pneumonia a number of times, and almost died. Her husband is a heavy smoker. Two of his brothers also smoke heavily. One has just been diagnosed with severe lung cancer, [and] has perhaps two months to live." [16]

Health Risks for Women

Many studies done in the past about the effects of tobacco use have been done on men and then generalized to women. However, more studies are being conducted on women smokers. These stud-

Studies on women smokers demonstrate that their risk of suffering a heart attack is 50 percent higher than that of male smokers.

ies show that women smokers have a 50 percent higher risk of heart attacks than male smokers. They are also more than twice as likely to experience depression and have a higher risk for osteoporosis and cardiovascular disease. Additionally, women smokers have a higher risk of lung cancer than men and a higher risk of cervical cancer than nonsmoking women. More studies, some on men and women and some on women alone, are being done to determine exactly how and why tobacco use causes these and other diseases.

One possibility for the differences between the sexes is that men and women have different hormones. Female hormones may affect the way tobacco products are processed in the body and blood. As an example, estrogen may promote cancer growth by promoting a binding effect between cancer-causing chemicals in tobacco and DNA in lungs. So, there are health effects from tobacco that are unique to women.

As well as harming women more than men, tobacco use during pregnancy creates additional risks for women and for their children. Nicotine and the other chemicals in tobacco and tobacco smoke can injure an unborn child in many different ways. Nicotine acts directly on the fetus's developing brain and reduces the amount of nutrients available. Instead of an unborn child receiving the rich, oxygenated blood that it needs, it receives blood full of carbon monoxide, hydrogen cyanide, and many other harmful chemicals, including nicotine.

Low birth weight is the most commonly cited problem with smoking while a woman is pregnant. Smoking during pregnancy has also been associated with miscarriages, stillbirths, longer stays in neonatal intensive care, higher incidences of sudden infant death syndrome, fussy or colicky babies, and learning disabilities. Because nicotine stimulates cells that release the neurotransmitter acetylcholine in the developing brain and nervous system, it can change the way a fetus's brain forms, permanently affecting anything from a child's personality to the ability to think. Problems with memory or thought processes (cognitive disorders) or learning disabilities may not show up until a child is in late elementary school or even adolescence. There may also be genetic cellular damage in the developing fetus that may put that child at a greater risk for cancer or other illnesses later. Many of these babies are born nicotine addicted, and after birth, when their nicotine supply is cut off, have to go through withdrawal symptoms.

Passive Smoking

Infants are not the only ones who are affected by the smoke of others. Millions of people expose families, friends, and those in their workplace to tobacco smoke on a daily basis.

Smokers inhale only a fraction of the smoke produced by a burning cigarette. The rest permeates the environment unfiltered and is breathed in passively by anyone nearby. This passive smoking is commonly referred to as environmental tobacco smoke (ETS), and has also been called secondhand smoke. The Environmental Protection Agency (EPA) classifies ETS as a class A carcinogen, the highest designation of a cancer-causing substance.

Some doctors have always recognized the risk of ETS. In fact, throughout history, physicians and scientists have reported that

Secondhand tobacco smoke also harms health.

people exposed to tobacco smoke suffer more illnesses and problems than those who are in smoke-free environments. In modern medical literature, passive smoking was first mentioned in 1974. Scientists reported that children in homes where there was ETS present had more respiratory diseases.

Children who smoke passively have an increased risk of asthma, ear infections, and respiratory tract infections like pneumonia and bronchitis. They also have decreased lung function. For example, when she was fourteen, Norah Teeter, the daughter of a smoker, stopped breathing from a severe asthma attack. After being resuscitated, the doctor told her mother that her smoking was affecting her daughter's health. Norah remembers that "My doctor told my mother she should stop smoking. I was under chronic abuse from the smoke. The smoke was a constant irritant and would increase the problem with asthma."[17]

Asthma and respiratory disease are not the only problems that are made worse by passive smoking. Other temporary health problems include headaches and eye, nose, and throat irritation. Because chemicals are released into the air largely unfiltered during the burning process, ETS can also cause permanent damage such as lung cancer and heart disease. In fact, over three thousand deaths each year are caused by passive smoking.

In the United States, ETS is a public-health issue. Many states have enacted laws that protect nonsmokers from being exposed to ETS in the workplace, restaurants, schools, and other public places. In many areas of the country, the tide has turned so much that smoking is considered socially unacceptable.

Although tobacco has earned a negative reputation in recent years and is becoming more and more socially unacceptable, a large percentage of the population is addicted to nicotine and continues to use tobacco on a daily basis. Despite the various health implications and public awareness surrounding nicotine addiction, many new people become tobacco users every day. Unfortunately, most of these new tobacco users are teenagers.

Chapter 3

Teens, Tobacco, and Trade-Offs

Tobacco companies know that young people who become addicted to tobacco may use it for their lifetimes. Because of that, they have targeted advertising to young people and teenagers. Since most adult tobacco users started as teens or younger, the focus of many medical professionals and public-health officials is to prevent young people from ever starting.

Despite the health warnings on cigarettes and all the new anti-smoking education and advertisements over three thousand young people start smoking every day. More than one thousand of those will become addicted to nicotine, and over 70 percent will regret ever having started before they were eighteen. Mona Vanek, a historian and writer, says, "[I] struggled to learn [to smoke] from 16 until I was 19, and then, for years and years I could only stand the awful taste if I was chewing mint gum. Boy, what a dummy I was! Wanting to fit in with my peers!"[18]

Young People and Tobacco

Imagine a high-school classroom full of thirty young people. At least eleven of them are using tobacco. Now imagine the same class in a middle school. At least four are using tobacco.

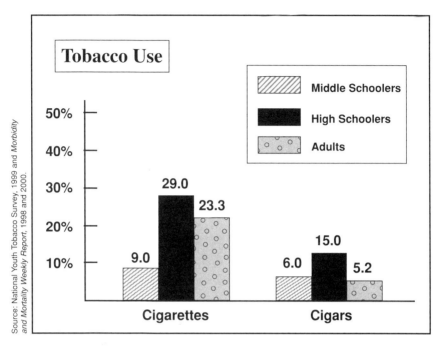

Source: National Youth Tobacco Survey, 1999 and *Morbidity and Mortality Weekly Report*, 1998 and 2000.

High-school students have traditionally used tobacco in much higher percentages than adults have. The 1999 National Youth Tobacco Survey stated that more than one-third (35 percent) of high-school students used some form of tobacco in the past thirty days. Twenty-nine percent smoked cigarettes, 15 percent used cigars, and almost 7 percent used smokeless tobacco. The use of smokeless tobacco products was almost exclusively by young men—12 percent used them versus only 1.5 percent of high-school females. Use of other cigarette products varied—5 percent of high-school students used bidis, and 6 percent used kreteks. These numbers were significantly higher (16 percent) among urban youth.

Middle-school students used tobacco products in roughly one-third to one-half the rates of high-school students. About 13 percent of middle-school students had used some form of tobacco in the past thirty days. Nine percent smoked cigarettes, 6 percent used cigars, 3 percent used smokeless tobacco products, and about 2 percent used pipe tobacco, bidis, or kreteks.

The Adolescent Adventure

Teenagers use tobacco at a much higher rate than adults do for a number of reasons. Erik Erickson, a psychoanalyst in the early part of the twentieth century, proposed that adolescence is a time for teenagers to establish their identity. During this time they try out a number of different identities—they may copy the language of someone they admire, try different styles of dress, or try a variety of different activities.

Barb Chandler, a freelance writer and psychotherapist, started smoking as an adolescent. She believes young people may start smoking as a way to create their own identities. She says,

> I think of myself at that age and realize how a person who smoked was considered "bohemian" or "worldly" and how I wanted to make this statement, so I smoked. Maybe that is why young people smoke—they want to make a statement and "march to the tune of a different drum." There are so many ways they can achieve their individuality today without having to risk their health and/or their lives.[19]

Some young people take up smoking as a means of establishing an identity.

Psychologists have identified a number of other needs that teenagers have. Teenagers need to take risks and rebel, to be respected and liked, fit in with their peers, express themselves, and establish their independence. Some teens believe that they appear older, more mature, or more independent when they smoke a cigarette. Gaining the approval of their friends or having their friends offer cigarettes can encourage them to become and remain smokers. Smoking can also help teens fit in with peers who smoke.

Becky Brooks started smoking at age nine and quit before her thirtieth birthday. She cites her parents as one of the influences on her decision to try smoking. She says, "Peer pressure was a secondary factor—every other kid, and they were older kids, who had an influence on me came from parents who smoked."[20]

Once young people have started to use tobacco, whether because of peers, or simply wanting to prove themselves, studies have shown that they may try other risky behaviors. Tobacco has often been called a "gateway" drug. A gateway drug is one that opens the gate to more extensive experimentation in drugs. Most states prohibit the sale of tobacco products to anyone under the age of eighteen as well as the use of tobacco by anyone who is underage. So young smokers are not only risking their health, they are taking a risk with the law. Risks can be exhilarating for anyone, and teenagers are no exception. It can be thrilling to do something illegal, and teens may want to prove wrong all the people who say tobacco is bad for their health. In fact, experts have found that teens engaging in risky behaviors like experimentation with drugs or sex are more likely to use tobacco regularly.

Using Tobacco to Cope

Many people, including teens, use tobacco as a way to deal with stress. Tobacco produces both stimulating and relaxing effects on the mind and body. Initially, a tobacco user may feel a burst of energy. But tobacco also acts like a tranquilizer or relaxant. During stressful circumstances, a stimulant can help provide someone the energy they need to deal with stressful situations. In addition, the relaxing effects help tobacco users keep their emotions under control,

Ten Ways Teens Can Cope with Stress

Janice P. Teeter has been a counselor for over twenty years. In an interview with the author, Teeter suggests ten coping skills that teens can try instead of using nicotine:

1. **Get enough sleep.** Adults need 8–9 hours of sleep every night. Teenagers need even more than that, yet usually get substantially less.

2. **Find a friend.** Talk about things with a friend, counselor, or family member.

3. **Write a diary.** Write down thoughts and feelings in journals. Emails or letters can be written, and then can either be sent or saved without sending.

4. **Take care of physical needs.** Eat veggies, and work out or take walks on a regular basis. Take extra walks during stressful times.

5. **Give yourself praise.** Think of accomplishments daily, and recognize interests, talents, and skills.

6. **Take some control.** Choose how to react to what happens or what others do.

7. **Make the stress "go away."** Problem-solve to find workable solutions for big stressors, such as homework (get a tutor, do homework at lunch so evenings are free, etc.).

8. **Recognize time limits on stress that can't change.** It can offer hope that some stressors may not change but will eventually end (e.g., after graduation).

9. **Deal with it.** Learn how to deal with or accept permanent stressors.

10. **Volunteer.** Take a few hours a week to tutor a child or help a new mother. Helping others can help put one's own problems into perspective.

and their thinking becomes clearer and more focused. This ability to be energized, calmed, and also have clearer thought processes helps tobacco users to cope with stressful difficulties they encounter.

Tobacco users experience withdrawal symptoms when the amount of nicotine in their body falls below the level of need, which can vary from person to person. These symptoms—irritability, anxiety, restlessness, and tension—can actually contribute to a tobacco user's perception of a situation as stressful. A smoker who had just smoked might be able to handle, for example, a pop quiz in school better than a smoker who felt he or she needed to smoke. Using tobacco increases the level of nicotine in the blood and thereby relieves the symptoms of withdrawal. Later, when the nicotine level decreases and withdrawal symptoms return, tobacco users turn to tobacco to alleviate their physical and emotional distress, thus creating an endless cycle in which nicotine is the controlling factor.

Risk Factors

Evidence shows that most people are under stress when they start using tobacco. Adolescence itself can be quite stressful, which could explain why there is a higher rate of tobacco use among teens than among adults. Many of the risk factors that lead to smoking, are, in fact, stressful situations. For example, studies have shown that childhood traumas can lead to smoking. Teens subjected to childhood physical, emotional, or sexual abuse are more likely to smoke. Other studies show that teens with family members with substance abuse problems or family members who have been in jail

Teens smoke for a variety of reasons, sometimes as a response to stress.

are at higher risk. Those whose parents are separated or divorced are also more likely to use tobacco. Additionally, those teens whose parents have mental illnesses, or those who have mental illnesses themselves, are more likely to begin using tobacco.

Besides these stress-related risk factors, studies have identified other teenagers most likely to become smokers. Males of any race have a higher risk; African American females have the lowest risk. There have been some studies which have indicated that children of smokers are much more likely to smoke. Other studies have indicated that this increased risk of tobacco use may be due to a hereditary predisposition to addiction. Yet others cite the environment the children of smokers are raised in. Babies whose mothers smoked during pregnancy are conditioned to respond to nicotine even before birth, and these children may be at higher risk of starting smoking as they get older. Nicotine also passes through breast milk, giving breast-fed babies of smoking mothers further exposure to the drug.

As children of smokers grow older, they may start smoking simply because it is familiar to them. Sometimes family members (parents, cousins, or older siblings) directly support a young person's smoking habit by purchasing the tobacco products for them. Or simply watching a mother, father, or grandparent smoke can be an additional risk factor. Approximately 9.8 percent of households have an adult smoker. Impressionable children admire virtually all aspects of their parents—even negative ones. Thomas Stanislao admired his grandfather, and it is this admiration that led to his twenty-year smoking habit. He tells of summer visits with his grandfather: "My grandfather had been a superb athlete . . . [and] a hero in WWI and WWII. I adored him. In the mornings, I liked to watch him shave—around the cigarette in his mouth. He chain-smoked unfiltered Chesterfields, probably three packs a day. He died at 77 of emphysema."[21]

Marketing to Young People

Young people exposed to tobacco advertising are more likely to start using tobacco products. The tobacco industry says it no

A cartoonist casts a sardonic eye on the marketing of cigarettes to ever-younger segments of the population.

longer markets to children or youth. Traditionally, however, tobacco companies have always looked to the youth market to increase their sales. They know, as scientists do, that over 90 percent of adults who smoke started when they were under the age of eighteen. They also know that they lose a number of adult smokers each year, primarily due to death or illness, and to maintain tobacco sales they must keep on marketing to gain new smokers.

In their marketing, tobacco sales representatives traditionally have paid special attention to cigarette outlets like convenience stores that are located near schools or places where young people congregate. They make certain that these outlets have better signage, prominent placement of tobacco items, and that they are well stocked at all times. One tobacco sales representative remembers that at an R.J. Reynolds company sales meeting, an employee asked the executive panel to clarify which young people were being targeted—junior-high kids or even younger children. The answer came back, "They got lips? We want 'em." [22]

Tobacco marketers understand the basic needs of teenagers. They hire psychologists and conduct research to keep current with teen interests and fads. They know that many teenagers want to establish their independence. So in an ad, they may portray a smoking man or woman who is independent and loving it. Since teenagers have a need to rebel, tobacco companies might show an ad of a maverick "thumbing his nose" at the establishment. Or an ad may portray a smoker as sophisticated, cool, or sexy.

Adolescents often believe the propaganda that the tobacco companies put out, especially when that propaganda is delivered by a cartoon. The most recognized cartoon character in the 1990s, even better known than Mickey Mouse, was Joe Camel. Joe Camel, a cartoon character used to sell Camel cigarettes, has gone into retirement. In 1998, tobacco companies were banned from using cartoon characters and billboards to advertise tobacco.

Camel used a cartoon character to advertise its cigarettes. Joe Camel was specifically designed to appeal to children and teens.

In addition, they are restricted from offering promotional items, since studies have shown that teens who own tobacco-company giveaways like jackets or visors are more likely to smoke.

Some ads that tobacco companies are running in magazines now tell young people not to smoke. However, at the bottom of those ads is the name of the tobacco company. Young people remember the names in the ads, and studies have shown name recognition is important to them—those who can name a brand of tobacco are more likely to use it. The most common cigarettes young people buy are brand-name cigarettes, whereas adults are more price conscious and are not as loyal to a particular brand. So the tobacco companies' antitobacco ads may work in reverse on teens—while the ads are not actively encouraging tobacco use, they are promoting brand- and company-name awareness.

Other Factors to Consider

There are enough challenges facing teenagers without adding the dangers of tobacco. Many aspects of health are affected when a teen uses tobacco. The chemicals in tobacco affect teenagers' bodies and minds, just as they do adults'. And teens are still growing and developing.

Studies have shown that teen smokers are clinically depressed at a higher rate than other teens. There is still debate on whether depression causes teens to smoke or whether teens who smoke become depressed. Recent studies have argued that it is the latter: Teens who smoke are at a higher risk than other teens for depression. In fact, smoking is one of the highest predictors of depression among teens.

Other studies have shown teens who smoke at least a pack a day are more likely to develop agoraphobia (fear of open spaces), anxiety disorder, and panic disorder. Another study cites not only depressive behavior in teens who smoke, but increased aggression as well. Sometimes tobacco use leads to decreased ability to perform in school. Additionally, when young people are trying to hide their illegal habit, they must be dishonest with both authorities at school and with their parents.

Many teens feel they cannot tell their parents about their tobacco use because their parents will not condone it. As a result, many teens end up sneaking cigarettes from others or stealing the money to buy cigarettes and other tobacco products. Laurie Teeter, who was introduced to smoking at age six by an older brother, says, "In order to get cigarettes, I would steal money from my mother. I would take money out of her purse and then go buy cigarettes. No one ever caught me. I was 12 or 13 and smoking up to a pack a day, usually after school." [23]

Prevention Efforts

Because of the lifelong effects and the difficulty in ridding themselves of this addiction, it is better for teens never to start using tobacco. Becky Brooks, who smoked for twenty years after starting at age nine, says, "Tobacco-related habits are dirty, smelly, expensive, and potentially destructive in a variety of other ways. . . . I think it is important to send a clear message to young people that it matters that they not smoke. I know both sides of this coin, life is better without the habit—and I loved my cigarettes. . . . Why not give ourselves a taste of better?" [24]

Some positive public policies have been put in place to discourage teenagers from using tobacco. Taxes increase prices, and higher prices may place tobacco out of the reach of many teenagers. William Evans, of the University of Maryland, analyzed surveys of teen smoking in a number of states. He says, "A 10 percent increase in the price of a pack of cigarettes will result in a 5 percent reduction in teen smoking. The evidence is overwhelming—higher taxes reduce teen smoking." [25]

Legal restrictions also help discourage smoking, and it helps even more if the laws are strictly enforced. Penalties for minors who are convicted for smoking vary from state to state—some states simply charge a fine, others require treatment programs. Amy Bloxham found out how serious it is to sell tobacco to minors after she lost her job because of it. She explains: "This guy looked like someone who regularly came in to purchase cigarettes. When I asked for ID, he said he had left it in the car. I said, 'Well,

I really shouldn't do this, but just this once.' Well I really *shouldn't* have done it. It was a federal sting, and this young man was a minor. I was fired immediately."[26] The store Bloxham worked for was also fined for her mistake, and will now be more closely monitored by the Food and Drug Administration (FDA), which oversees tobacco sales.

Besides strict enforcement of the ban on tobacco sales to those who are underage, it helps to have restrictions and strict antismoking policies at home, at school, in places of employment, and in public places. Many restaurants and the majority of public places in the United States prohibit smoking. A majority of workplaces also restrict smoking in areas where it affects other employees or customers. The Boeing Company, which manufactures commercial and military aircraft among other products, restricts employees from smoking within twenty-five feet of the doorway of any building. This can be difficult for employees because Boeing factories are among the largest buildings in the world. Factory workers have ten-minute breaks, and much of that time can be spent walking in and out of the buildings to smoke. Many other companies have similar policies. Companies that have strict smoking policies often provide free or low-cost smoking-cessation programs.

Most schools recognize that smoking is illegal for minors. Some schools have very strict restrictions on smoking, and students caught smoking are suspended from school. Other schools will not allow students who are caught smoking to participate in extracurricular activities or sports. In some states, such as Oregon, the use of tobacco by minors is a misdemeanor crime, and schools must intervene with some sort of smoking-cessation program if students are caught.

School-based prevention programs can provide education to teens on the negative health effects, social consequences, and costs of tobacco use. In addition to education, counseling and intervention with young people most at risk can stop teens from using tobacco before they become addicted. Since most young people attend schools, schools are in a unique position to counsel, teach, and help them with this important health issue. School-based programs are especially effective in reducing the rate of smoking initiation when used in combination with community programs such as educating store clerks, involving parents, and using the media to provide antitobacco messages.

A survey of over seventeen thousand teenagers found that with increased restrictions, teens believed that smoking was inconvenient and often socially unacceptable as well. Myra Nelson, who has been a nonsmoker for five years, learned about the social consequences of being a smoker. She says, "Social prohibition was what made me quit. I began feeling like a social leper. Smoking became embarrassing."[27] As young people become more educated about the consequences of tobacco use, they start making better choices overall.

SWAT

A group of Florida youth decided to take tobacco education into their own hands. They started their own media campaign with money from the 1998 tobacco settlement, in which tobacco com-

Two teen members of SWAT use their cell phones in a campaign against smoking.

Sean Marsee

Sean Marsee, of Ada, Oklahoma, was one of Talihina High School's most outstanding athletes. He had won twenty-eight track medals. However, as many races as he won, he lost the most important battle of all. He lost a ten-month battle against cancer.

Sean had been using moist snuff since he was twelve. But besides his tobacco habit, he took care of his body. He ran five miles a day, ate well, and lifted weights. When he was eighteen, he noticed a red spot on his tongue. The doctor took a biopsy, and the diagnosis was cancer.

Eventually Sean lost most of his cheek and his jawbone to the cancer. After three surgeries, radiation, and chemotherapy, the cancer finally won. Sean died quietly at home. He was nineteen years old.

---

panies agreed to pay money to compensate states for the expenses of sick smokers. Students Working Against Tobacco, or SWAT, was started, and the Florida teens developed an antismoking campaign directed at the reasons that young people smoked. Studies have shown that teens who regularly watch antismoking messages are half as likely to start smoking as other teens.

The Florida teens created an ad telling young people that the tobacco companies were manipulating facts and manipulating them with advertising. Instead of smoking to be rebellious against the establishment or their parents, teens should not smoke because by smoking they are conforming to the mass brainwashing and manipulation of the tobacco industry.

SWAT created a number of ads and a website (www.whole truth.com). Visitors to the website can play a game to show them how the tobacco industry manipulates them. They discover for themselves how tobacco ads address every one of the issues that teenagers often struggle with. Instead, SWAT seeks to tell teens the truth: Smoking will not solve problems, help young people fit in with their peers, or magically make the pain of adolescence go away. It simply causes more problems for the present and in the future.

In its first two years of operation, SWAT managed to convince thousands of teens not to smoke that first cigarette. In a study done in 2000, the number of Florida middle-school students who used tobacco declined from 15 percent in 1999 to 9.6 percent in 2000. The number of high-school students who used tobacco declined from 26.2 percent in 1999 (after one year of SWAT's operations) to 20.9 percent in 2000. This is well below the national average of 35 percent. SWAT's successful antitobacco advertising campaign led officials to believe that implementing statewide programs could be effective in preventing and reducing tobacco use among teens.

Never Starting

The message of SWAT and others is clear. The first cigarette causes damage. Nicotine in the first cigarette can cause addiction. It is very important for teens never to smoke the first cigarette. Jennifer Nelson, a freelance writer, says,

> I've spent most of my adult life addicted to nicotine. . . . My kids have seen my husband's and my struggle to quit-and-stay-quit and they lost their grandmother at a much too early age. I am fairly confident I've conveyed the fact that never starting is the only way to make sure that they will not have to spend their entire adult lives "beating" the addiction. I'm pretty sure my two won't ever light up. But if every young person could "get that" by age 16 or 17, the odds would be very favorable for keeping them smoke-free.[28]

Never starting to use tobacco products may not be easy for some teens. Young people may have to defend their choice not to smoke or use other tobacco products to their peers. It is hard for young people to do things differently from friends. It is much easier to give in to peer pressure than to give in to the subsequent addiction. To many teens the negative consequences related to tobacco use seem too far into the future to be a reality.

Yet, it is important to make a stand against tobacco use, because the alternatives are some very difficult fights—the fight for good health, and potentially, the fight to quit. Once someone has started using tobacco, nicotine will keep him or her addicted. If that person ever decides to quit, he or she will make an average of five to seven quit attempts before they are successful. The struggle to quit can literally encompass an entire lifetime.

Chapter 4

Nixing Nicotine

There are many benefits to quitting tobacco use, and even more to never starting. Almost 90 percent of smokers who started smoking as teenagers wished they had never begun. Even more wish they knew an effective way to quit. But it is not easy to quit permanently. In fact, author Mark Twain purportedly once said that quitting smoking was easy—he had done it hundreds of times.

There are many different resources available as well as a variety of methods to help people quit using tobacco. For instance, quitting smoking depends on the smoker. No one method works for every person because the situations that trigger their smoking are different. Even the reasons smokers want to quit can differ.

Many programs are targeted for the age or characteristics of the specific tobacco user. Some methods work well for adults but do not work as well for teenagers. Teens have very different needs than adults, and women have different needs than men. Cessation programs should be customized for the gender and age of the tobacco user. Usually, a combination of methods and approaches works better than a single method. The methods chosen, combined with a very strong desire and commitment, will help a tobacco user quit. Madeleine Armstrong, who has been a nonsmoker for fifteen years, says,

> If I could quit, anyone can. There is a secret. I will share the secret with you because I feel that smoking is so detrimental to one's health and if anyone wants to quit, they can. . . .

> The secret? The secret is that you have to want to. That's all there is to it. You really, really, really, really, really have to want to! You have to manifest what you believe. You have to change your way of thinking. You have to change what you are doing. You have to change the way you feel about why you are smoking. You have to become a witness of your life.[29]

After Quitting

Quitting tobacco can cause a number of changes in the body. Some people wonder if quitting would be of any benefit, since their bodies have already been damaged from tobacco use. However, there are a number of changes that happen immediately, and after time, some of the damage is reversed.

Twenty minutes after the last tobacco use, body temperature in extremities, the pulse, and blood pressure return to normal. After

The Benefits of Quitting

20 minutes	After quitting, a smoker's blood pressure and pulse rate reduce to normal, as does the body temperature of his or her feet and hands.
8 hours	After quitting, carbon monoxide level drops and oxygen level rises to normal.
24 hours	After quitting, the risk of heart attack decreases.
48 hours	After quitting, a smoker's ability to taste and smell is enhanced.
2 weeks	After quitting, circulation improves, walking is easier, and breathing efficiency increases by nearly 30 percent.
1 year	After quitting, the risk of coronary disease declines by 50 percent.

eight hours, oxygen and carbon monoxide in the blood return to normal levels. The chance of heart attack diminishes in twenty-four hours. After two days, smell and taste start to return to normal. After three days, shortness of breath decreases and lung capacity increases. Within three months, circulation improves, lung function improves dramatically, and cholesterol levels plummet. Within nine months, energy levels increase and coughing and lung infections decrease.

After one year, the risk of coronary heart disease for former smokers is half that of smokers. Within two years, the risk of heart attack nears that of nonsmokers. After five years, stroke risk is reduced and risks of lung, mouth, throat, and esophageal cancers is half that of smokers. Lung cancer risk is near that of nonsmokers after ten years, and, after fifteen, the risk of coronary heart disease is near that of nonsmokers.

So quitting, even after long-term usage, can help former tobacco users become healthier. It is for precisely that reason that many people choose to quit.

Quitting Cold Turkey

To quit cold turkey is to stop using tobacco abruptly and with little or no preparation. Kathleen Purcell, a newspaper editor, says "My father quit 40 years ago when he fell asleep and awoke to find the chair he was sitting in was in flames. He quit cold turkey." [30]

People have a number of explanations about the origin of the phrase cold turkey. Some say that in the state of drug withdrawal the blood is directed toward vital organs, often giving addicts goose bumps. Their skin resembles that of a cold, plucked turkey. Others attribute this phrase to the idea that cold turkey is a food that needs very little preparation.

Whatever the origin of the phrase, quitting tobacco cold turkey does not work well. Studies have shown that less than 5 percent of smokers who quit in this manner are able to remain smoke-free after a year, but smokers who have a compelling reason to quit, such as a severe illness, have more success with the method. Jodi Waxman was diagnosed with lymphangioleiomyomatosis (or

LAM), a serious and often-fatal lung disease often initially misdi-agnosed as asthma or emphysema. LAM affects young women of childbearing years. Abnormal muscle cells grow in the lungs, clos-ing off the air space, and causing lung collapses and other serious complications. Many LAM patients eventually find themselves on a lung transplant list. Waxman says, "I quit smoking last February when my doctor suspected that I had LAM. I quit cold turkey be-cause she told me that smoking limited my lung capacity more than it already was."[31]

Although quitting cold turkey can work, people often start using tobacco again. One of the most common reasons that people who quit cold turkey gave for going back to tobacco was the intense nicotine craving, an almost irresistible urge to use tobacco. These cravings, combined with withdrawal symptoms like irritability and depression, can make it more likely for former tobacco users to re-sume their habits. A combination of quitting cold turkey and nico-tine replacement therapy (NRT) can help make the cravings more manageable and allow these people to remain tobacco free.

Nicotine Replacement Therapy and Behavior Modification

NRT usually consists of some means other than the use of tobacco to deliver nicotine to the body. Nicotine can be added to chewing gum, delivered into the bloodstream by means of a patch affixed to the skin, a nasal spray, or delivered in an inhaler similar to a cig-arette. Colleen Brady, a pharmacist, said in an article for *Chate-laine,* a Canadian women's magazine: "Nicotine replacement therapies work by providing the brain with a controlled amount of nicotine either through the mouth or the skin. The amount of gum chewed and the strength of patch used is slowly decreased over a period of time, about three months for the gum and eight to 12 weeks for the patch."[32] During all types of NRT, the amount of nicotine in the replacement product is gradually reduced until it is completely eliminated.

People using NRT are trying to minimize the effects of sudden withdrawal from nicotine. During nicotine withdrawal, studies

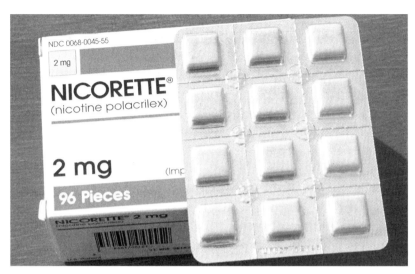

Gum containing nicotine is one of several products designed to help tobacco users kick their habit.

have shown that there are specific mental functions that are impaired. Although there is no change to complex cognitive functioning like logical reasoning, people who are withdrawing from nicotine often have problems with tasks requiring sustained attention. The patch or gum can help them gradually get used to changes in thinking processes. Furthermore, nicotine replacement can help the user focus on the behavioral and coping aspects of quitting tobacco without having to deal with the immediate problem of nicotine addiction or withdrawal.

People who want to quit smoking have to deal not only with the addictive nature of the drug but with their habits as well. Sometimes they will change their habits entirely, such as avoiding places in which they used to smoke, in order to break the habitual aspects of smoking. Smokers often want to have something in their hands, or even in their mouths, so people quitting smoking are often advised to chew cinnamon sticks or carry raw vegetables around as a substitute for cigarettes.

Being able to make decisions and be firm about them helps. Judith Stock, a freelance writer and editor, has been a nonsmoker for one year. When she quit, she took decisive action and removed all smoking-related items from her home. She says, "I just started thinking of myself as a nonsmoker as opposed to a smoker. I took all the ashtrays out of the house, tossed out the lighters. Everything that had anything to do with smoking, I tossed out. When I wanted to smoke I would take a walk around the block."[33]

A combination of NRT and behavioral modification techniques can be very helpful for many adults. However, several studies have shown that NRT is not an effective treatment for adolescents. In one study, which used a combination of minimal behavior modification and NRT, smoking abstinence was only 10.9 percent at six weeks and 5 percent at six months. This is not much better than the success rate of quitting smoking cold turkey.

Those teens who used NRT and started smoking again said it was because of stressful situations in their lives or pressure from their peers. NRT helps with the cravings so the person quitting can focus on other issues, but it does not eliminate those other issues. Other techniques may be needed to address those problems.

Hypnosis

One technique used by many people to quit using tobacco is hypnosis. Some people think hypnosis is a type of hocus-pocus or a way to control someone's mind. Actually, hypnosis is a state of concentration. If someone is intensely watching a television show, he or she can be so focused on the program that the outside world seems to "go away." A hypnotized person is in the same state—a state of intense concentration.

During that state of concentration, patients can look inside themselves and find, for example, what voids tobacco is filling. If, for example, a person was using nicotine to fill a void left by an absent parent, that person could use hypnosis to help heal feelings of emptiness. Using guided imagery techniques, that person could imagine feeling loved and wanted—by a parent, by a higher power or God, or by a friend. Using that image or experience, a person can reimmerse him- or herself in that feeling of being loved at any time and use those positive feelings to help overcome nicotine cravings.

Hypnosis helps some people quit smoking by improving their coping techniques.

Hypnosis can help people deal with other reasons they use tobacco. It can help establish better coping techniques and can help people learn to fill other types of voids, such as not feeling accepted, appreciated, or liked at school. In addition, since hypnosis is a heightened state of concentration it can help people stop ignoring the negative physical effects of the tobacco in the body—the lightheadedness, the searing lungs. Essentially hypnosis may make someone more cognizant of the effects happening in his or her body every time nicotine is taken into the body—which can lead to a desire to stop using tobacco.

Aversion Therapy

Some people use aversion therapy to quit smoking. Aversion is avoiding a thing or a behavior because it has been associated with something disgusting, unpleasant, or painful. Aversion therapy helps people learn to associate things they wish to avoid with a feeling of disgust or physical and emotional discomfort. Because of this negative association, those using aversion therapy to quit smoking modify their behavior and avoid tobacco products.

Imagining all the terrible effects tobacco has on the body is a way of using aversion therapy to stop smoking. When smokers think back to the first time they started cigarettes, they can recall the nausea, dizziness, and other negative effects they may have experienced and imagine they are experiencing those discomforts again during each cigarette. If this is imagined often enough, smokers tend to associate the cigarette with negative feelings and will not want to smoke.

Another way of using aversion therapy is to imagine that the cigarette is something physically disgusting. For instance, a smoker might imagine that the cigarette is filled with bugs or worms instead of tobacco. The guided imagery of hypnosis can greatly enhance aversion therapy as well. Because someone under intense concentration can often create vivid mental images, there can be a strong association between the negative images and the use of tobacco.

Smoking-Cessation Groups, Mentors, and Community Help

While hypnotists and aversion therapists help smokers with specific techniques to quit smoking, smoking-cessation groups and support systems help people help themselves. It can be helpful to have a friend who has had similar experiences. Smoking-cessation groups might be offered through schools, hospitals, and churches. These groups provide education and guide new nonsmokers by helping them through the rough times. Many smoking-cessation programs also teach methods to manage anger and stress and teach ex-smokers how to keep a handle on their emotions. This can be critical because many people use tobacco to cope with anger or stress, and when they stop using it they need to learn new methods for coping.

Mentors are people who teach and guide other people, often through a difficult process. They can provide insight, inspiration, and aid to the struggling ex-smoker. For example, mentors may make themselves available to talk whenever someone experiencing

The smoking habit is a powerful addiction, and quitting can be extremely difficult.

intense cravings is worried about relapsing. Mentors can be contacted in person or online. Some mentors guide people one-on-one, others offer e-mail mailing lists through which they assist many smokers or soon-to-be-ex-smokers at once.

Nicotine Anonymous is modeled after the highly successful Alcoholics Anonymous (AA) program. Participants meet with others struggling to rid themselves of nicotine in their lives. They share experiences, strength, and hope with each other. Utilizing twelve-step programs originally developed for AA, participants help each other on a path toward achieving personal responsibility and remaining nicotine free. Some Nicotine Anonymous groups have a religious character; others do not.

Religious organizations can also provide a network of support for the smoker, who may be going through one of the most difficult times in his or her life when quitting. Scientists have found faith to be helpful in a number of health situations. Meditation or prayer can help people with the urges and cravings they experience. Dave Waddle, a contractor, describes his experience in praying for help in quitting smoking. "I said a prayer, and that was five days before my son was born. My dad smoked for years and died recently of lung cancer. I didn't want my son to see me that way, lying in a hospital bed, slobbering on myself. I held my dad's hand as he took his last breath. After my prayer, the urge for smoking went away. My son is a month old now." [34]

Medical Help

Besides finding help from religious organizations, people wanting to quit using tobacco may seek help from physicians and other medical practitioners. Physicians often talk to smokers at yearly physicals or when they are ill. They usually counsel patients on the health risks of smoking. Large health maintenance organizations (HMOs) like Kaiser Permanente or Group Health Cooperative of Puget Sound offer smoking-cessation programs. Some programs like this are even free to the consumer—the HMOs believe that it is cheaper to pay for a program to stop smoking than it is to pay for the treatment of many of the diseases and the complications that can arise from a lifetime of nicotine abuse.

Physicians can offer some treatments that are only available with prescriptions. Bupropion is a medication (Zyban) authorized by the FDA to be used for smoking cessation. The same medication is also marketed as an antidepressant under the brand name Wellbutrin. "Bupropion helps curtail the cravings," says pharmacist Frederick Nehser. "It helps take away the need [for nicotine.]" [35] Medications such as clonidine and nortriptyline can help with the side effects of withdrawal, but bupropion is the most commonly prescribed medication because it alters the brain chemistry so the nicotine cravings are reduced.

Scientists and researchers concerned with nicotine addiction are regularly looking for new ways to help people quit or to help prevent them from becoming addicted in the first place. Scientists are exploring medication and vaccines that help block the action of nicotine in the brain. If nicotine cannot bind with the receptors in the brain responsible for addiction and pleasure, then the smoker will not experience the pleasurable effects of the nicotine. When the internal reward for tobacco use is gone, smokers will find little reason to continue.

Facts About Quitting

When people are first seeking to quit, they most often try medication or NRT in addition to either a support group or behavior therapy. This combination of counseling and NRT is probably the most effective in helping people quit and stay smoke free for the long term. Many existing therapies are more effective when used in combination with other therapies. One study showed that the success rates (3 to 5 percent) for cold turkey double for those receiving brief advice and triple for those receiving behavioral therapy. Hypnosis, aversion therapy, and smoking-cessation groups generally average success rates of about 10 to 15 percent. Most of these therapies work to modify behavior and have success rates similar to traditional behavioral therapy and counseling. However, teens often drop out of group treatment programs, perhaps because they are fearful of having to disclose to their parents that they smoke. Some treatment programs require parental consent. Groups that assure anonymity can be more effective in helping teens stop using tobacco, as can online help and individual mentoring.

For those tobacco users receiving medication, 10 percent were able to quit. Twenty percent of those receiving medication and

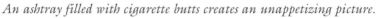

An ashtray filled with cigarette butts creates an unappetizing picture.

Weight Gain

Nicotine affects the appetite. There is some truth to the idea that smokers can control their weight with nicotine. This effect happens for a number of different reasons. First, smoking impairs taste and smell. With impaired taste, appetite can be diminished. Second, nicotine use can speed up metabolism, so the body works harder and burns more calories in general. Third, nicotine itself decreases the appetite. In fact, the chemicals in tobacco interfere with the absorption of many food substances and nutrients. This makes smokers more likely to have disorders such as osteoporosis, since tobacco interferes with the absorption of calcium.

Tobacco users who want to quit should expect that they might gain weight. Most who quit gain just a few pounds, which disappear relatively soon. About 10 percent of quitters gain more weight, which tends to be more difficult to take off. However, the negative health effects of excess weight are less than the negative health effects of tobacco. If someone has to make a choice between being overweight or smoking, it is often better to be a few pounds too heavy.

One way to help with the weight issue is to make certain the diet is wholesome by eating fruits and vegetables, rice, millet, or potatoes without toppings; reducing the amount of meats and fatty, refined, or processed foods; and snacking on carrots, celery sticks, or unsalted seeds and nuts. Other ways to control weight gain include joining a health club, starting an exercise program, or getting a dog. Studies have shown that dog owners who walk their dogs regularly walk more than other pet owners and are in better health and condition overall.

It is not uncommon to gain weight when quitting smoking; exercise can help control weight gain.

brief advice quit, as did 30 percent of those receiving medication and therapy. The success rate with NRT doubles when combined with counseling or behavioral modification.

The success rates for NRT vary depending on the type of nicotine used. In a 1999 article in the *Journal of the American Medical Association,* it was reported that people using the nicotine patch alone have a quit rate of 5 to 11 percent. Those using nicotine gum have a success rate of 13 to 15 percent. The nicotine inhaler, which uses a cigarette holder attached to a nicotine cartridge, has a success rate of 17 to 21 percent. The nicotine nasal spray has a success rate of 25 to 35 percent. Some recent studies have indicated that a combination of nicotine replacement strategies also improves the chance for success. Success rates are higher when combining the patch, which delivers a steady dose of nicotine, with nasal spray, which provides bursts of nicotine that move quickly into the bloodstream and the brain.

However, these success rates are just percentages and do not indicate which method will be most successful for any given tobacco user. Some people will try a number of different methods before finding one that works for them. The truth is that quitting tobacco use is difficult. As with most other addiction-cessation programs, the overall success rates of long-term abstinence from nicotine are low. Only half of those who attempt to quit smoking succeed for more than two days. Most people who successfully quit make five to seven attempts before they succeed. Jennifer Nelson, a nonsmoker for two years, says, "This last time [I quit] I used the patch. I found it very effective in taking the edge off those first shaky few weeks of withdrawal. This is my sixth time." [36]

Once Addicted, Always Addicted?

Quitting tobacco use is very emotional. The nicotine has changed the way the brain works with every dose. Its loss makes the tobacco user feel incomplete. As a result, some people experience intense mood drops. In an article that examines the current and potential state of cessation therapy, one new nonsmoker said: "It was like losing a buddy. There was a real loss for me, an emotional loss. I probably cried every day the first week. I wanted to smoke."[37] The depression or mood drops may be because of an emotional or chemical withdrawal from the cigarettes or tobacco use. Depending on the severity of these depressions, some people may be inclined to start up again. In fact, research has shown that if there are severe mood disturbances within five days, the quit attempt will probably fail. Most people who are affected this way will resume their tobacco use within two weeks. Antidepressants are sometimes used as short-term therapy to help tobacco users weather this period of time.

Some people experience this emotional withdrawal immediately after quitting, and then again after some time has passed. Often after a month or two, the ex-smoker will experience an acute desire for nicotine and another serious mood drop. This, too, can put the quit attempt in jeopardy. Thomas Stanislao, a nonsmoker for six years, says,

> Five months [after I quit], I was at a conference in Palm Springs and had a little too much bourbon. Several people were smoking and I bummed one, expecting it to taste awful, to make me dizzy. It tasted wonderful, and I bummed another one. Then I thought, once an addict, always an addict. I know a woman who quit for 16 years and then started again. I was determined that would not be me. I've been clean ever since and never even think about it.[38]

The longing for nicotine often does not go away with time. The ex-user is still subject to nicotine addiction even years after quitting. Ten or fifteen years after people quit, they may still desire the chemical and to feel it acting in their brains. However, it does frequently lessen and people who experience the desire learn how to cope with the urges without using nicotine. Sometimes the desire can be enhanced by circumstances—when other people are smoking or the ex-smoker is in a situation in which he or she used to smoke, the desire may seem to be overwhelming. Once someone is addicted to nicotine, they are always addicted. However, just because the addiction remains does not mean they have to give in to that desire.

Tobacco is legal because it was in use centuries before the birth of the U.S. government. But perhaps if tobacco were discovered

A Five-Day Plan for Quitting Tobacco Use

The U.S. surgeon general's office publishes a five-day plan for quitting to-bacco use on their website, available at www.surgeongeneral.gov. They recommend that first, a person must decide to quit. Then they should contact a health-care provider or a tobacco-cessation clinic to discuss options. And before starting the plan, a person should set a quit day.

Quit Day Minus 5: List all of your reasons for quitting and tell your friends and family about your plan. Stop buying cartons of cigarettes or pouches of tobacco.

Quit Day Minus 4: Pay attention to when and why you use tobacco. Think of habits or routines you may want to change. Make a list to use when you quit. Think of new ways to relax or things to hold in your hand instead of a cigarette.

Quit Day Minus 3: Make a list of the things you could do with the extra money you will save. Think of who to call when you need help, like a smoking support group.

Quit Day Minus 2: Buy the over-the-counter nicotine patch or nicotine gum, or get a prescription for the nicotine inhaler, nasal spray, or the non-nicotine pill, bupropion SR. Clean your clothes to get rid of the smell of cigarette smoke.

Quit Day Minus 1: Think of a reward you will get yourself after you quit. Make an appointment with your dentist to have your teeth cleaned. At the end of the day, throw away all matches, cigarettes, and other tobacco products. Put away lighters and ashtrays.

Quit Day: Keep very busy. Change your routine, and do things differently so you're not reminded of tobacco. Let family, friends, and coworkers know this is your quit day, and ask them for help and support. Avoid alcohol. Celebrate, or buy a treat for yourself.

Quit Day Plus 1: Congratulate yourself. When cravings hit, do something not connected with tobacco, such as walking, drinking water, or taking deep breaths. Call your support network. Find snacks like carrots, sugarless gum, or air-popped popcorn.

today, it would be regulated as strictly as any other type of potent pharmaceutical. Perhaps it would never make it onto the market because of its negative health impacts. In the last century, particularly the time from 1964 on, the U.S. government has taken an increasing role in the relationships between the consumer, medical experts, and the tobacco industry.

Chapter 5

Advertising, Big Business, and Uncle Sam

When the negative health effects of tobacco use were first made public, the medical community pressed the government to respond. Through the last half of the twentieth century, the U.S. and state governments have increased regulations and taxes on tobacco products, regulated advertising, and sued tobacco companies for the medical costs that have resulted from smoking-related illnesses. The result has been a dramatic change in tobacco use in the United States.

The "Healthful" Cigarettes of the 1950s

Although scientists have published studies through the years showing the negative health effects of tobacco use, most of these have been ignored or downplayed by the media and the tobacco industry. Often the tobacco companies responded by publishing their own information touting the health benefits of tobacco use. Or they would propose other theories, including some that stated that viruses were responsible for lung cancer. Finally, in 1950, The *Journal of the American Medical Association* published studies that definitively linked smoking with lung cancer. These reports scared the public, and cigarette sales fell.

In the 1950s, tobacco companies introduced filtered cigarettes to combat studies linking smoking with lung cancer.

The tobacco companies recovered business by making changes to their cigarettes. One change was the introduction of special filters that were supposed to reduce the amount of tar and nicotine the smoker would inhale, and thus decrease the hazards to the tobacco user. Lorillard introduced a special Micronite filter, which purportedly provided "the greatest health protection in cigarette history."[39] Its secret ingredient? Asbestos, a substance now known to cause cancer itself. (Lorillard discontinued the use of this filter in 1956.)

In actuality, there was little difference between the amounts of tar and nicotine in filtered and unfiltered cigarettes, but the public believed the tobacco companies' ads. At the beginning of the fifties, 2 percent of cigarettes were filtered. By the end of the decade, over 50 percent were.

Men in white lab coats, who looked like doctors, were employed by the tobacco companies to deliver advertisements in support of filtered cigarettes. Liggett and Myers advertised the results of their own tests in 1952, which supposedly demonstrated that "smoking Chesterfields would have no adverse effects on the throat, sinuses or affected organs."[40] In 1954, Liggett and Meyers ran ads that featured the popular actresses Barbara Stanwyck and Rosalind Russell. They called their new alpha-cellulose filter "just what the doctor ordered."[41]

Cigarette sales rebounded with the tobacco industry's reassurances to the public. However, the government did not like all the unfounded health claims. In 1955, it clamped down. The Federal Trade Commission (FTC) published new rules that prohibited tobacco advertisements from referring to the throat, larynx, lungs, nose, or other parts of the body, and banned advertisements pertaining to digestion, energy, nerves, or doctors. With these new rules, the FTC hoped to avoid unsubstantiated claims that tobacco use improved health. However, there were still many unanswered questions about what role, positive or negative, tobacco actually did play in human health.

The 1964 U.S. Surgeon General's Report

In the early 1960s, various health organizations, the American Cancer Society, and other groups pressured President John Kennedy to find out definitively whether tobacco use was harmful or not. In 1962, a committee of eminent scientists and experts was formed to study the issue. Headed by U.S. surgeon general Luther Terry, a heavy smoker from the South who had picked tobacco as a boy, the committee was chosen very carefully. Terry gave the tobacco industry the right to veto any committee member so that they could not later claim the committee was biased.

Half of the ten committee members used tobacco themselves. All the tobacco users on the advisory committee agreed not to alter their smoking habits during their review so that they would not appear to be endorsing the view that tobacco was harmful. The committee's work would be kept entirely secret from the general public.

As U.S. surgeon general in the early 1960s, Luther Terry (pictured with Senator Robert Kennedy, right) headed a committee to study whether tobacco use was harmful.

Once formed, the committee examined data and statistics from over seven thousand studies published over the years. They visited medical researchers to look at their slides of pathological changes in smokers' lungs and examined scientists' work in the field. They listened to researchers who told them their views on the dangers of tobacco and the scientific basis for those views. For instance, one view stated that there were two genetically different types of people. If one type of person was genetically predisposed to tobacco use, perhaps this same genotype was also genetically predisposed to lung cancer. This could mean that any apparent link of smoking to lung cancer was related only to a person's genetics and not to a cause-effect relationship.

In 1964, after two years of study, the committee finally revealed its conclusions. The report announced that smoking causes lung cancer and laryngeal cancer in men, was a probable cause of lung cancer in women, may cause heart disease, and was the most important cause of chronic bronchitis. In addition, the committee discovered a relationship between smoking and emphysema and found that cancer-causing compounds (seven at the time) were present in cigarettes. The committee called tobacco "habit forming" rather than "addicting." (A habit can be changed more easily than an addiction—nicotine was first called addicting in the 1988 U.S. surgeon general's report.)

The report also noted that using tobacco had the positive benefit of promoting intestinal health and countering obesity. However, those benefits did very little to temper the message: Cigarette smoking was a serious health hazard and a significant threat to public health in the United States.

The extremely negative character of the 1964 report stunned the tobacco industry. They had believed that the surgeon general's report would be more balanced—if not in their favor, at least not so overwhelmingly against them. They immediately tried to find fault with the report and did point out several potential problems. One complaint was that highly filtered cigarettes had not been on the market long enough for researchers to accurately determine mortality rates of

Surgeon General Luther Terry (center) and two fellow committee members present their long-awaited 1964 report detailing the hazards of tobacco use.

smokers using them. However, despite that and other complaints, the body of the report from the Surgeon General's Advisory Committee stood up to public scrutiny. Smoking rates in the United States dropped almost 20 percent soon after the report came out.

Warning Labels and the Fairness Doctrine

Just one week after the surgeon general's report came out, the FTC announced that warning labels would be required on all cigarette packs in the future. In 1965 Congress passed a law that required the statement "Warning: Cigarette Smoking May Be Hazardous to Your Health" to be written on all cigarette packs. (These warnings have since gotten more detailed, and in Canada, warning labels actually display pictures; for example, a diseased lung or a diseased mouth.)

Although hotly debated, it was decided that warnings would not be required in advertisements at that time. This was in part because the tobacco industry had voluntarily restricted tobacco ads from mentioning any implied or actual health benefits from tobacco in general. An industry-selected monitor reviewed and rejected ads that showed too many positive aspects of cigarettes and tobacco. Yet tobacco companies were not restricted from showing their products as being pleasurable and having taste and flavor. As a result, tobacco companies purchased quite a bit of advertising time to tout the benefits of their brands.

However, the Fairness Doctrine required television companies to give equal time to opposing viewpoints on any subject. In 1967, John Banzhoff, a young lawyer who believed the Fairness Doctrine applied to tobacco advertising, petitioned the FTC. Although in news stories the broadcast companies presented both sides of the argument, Banzhoff argued that the commercials sponsored by tobacco companies were one-sided and did not present the opposing viewpoint that tobacco is unhealthy. Banzhoff took his case all the way to the Supreme Court and won. By 1969, broadcast companies were required to give free, equal advertising time to antitobacco groups. Ads created by the American Cancer Society and others did much to reduce smoking rates

Some Important Events in the History of Environmental Tobacco Smoke

In the last three decades, public policy has shifted toward the non-smoker. Most workplaces and public buildings are now smoke free, as are buses, trains, and planes. Restaurants are either smoke free or have ventilated smoking sections. This excerpt entitled "Tobacco Timeline" from the Tobacco BBS (www.tobacco.org) shows important events in the history of environmental tobacco smoke.

1969   Ralph Nader asks FAA to prohibit in-flight smoking since it is unhealthy and a fire hazard

1969   Pan American creates the first nonsmoking section

1972   The US Surgeon General's Report talks about secondhand smoke or "public exposure to air pollution from tobacco smoke"

1983   San Francisco bans smoking in private workplaces

1987   Congress prohibits smoking on short flights (less than two hours)

1987   Beverly Hills, CA, bans smoking in restaurants

1988   New York City passes "Clean Indoor Air Act"

1990   San Luis Obispo, CA, becomes the first city to ban smoking in all public buildings, including restaurants and bars

1990   Bill Clinton outlaws smoking in the White House

1990   Smoking not allowed on most domestic flights of less than six hours

1990   Smoking prohibited on interstate buses

1993   Vermont is the first state to ban indoor smoking

1993   US Postal Service bans smoking in its facilities

1994   Restricted smoking at US military bases

1994   McDonald's bans smoking in all of its restaurants

1995   New York City makes stronger laws with the "Smoke-free Air Act"

1998   California bans smoking in bars

2002   Salt Lake City Olympics maintains a smoke-free policy for athletes and spectators

No Smoking signs have become increasingly common in the last three decades.

by presenting the health information and known facts about tobacco usage to the American public.

In 1971, the tobacco industry voluntarily pulled all cigarette ads from television and radio in exchange for Congress's decision to delay controls and restrictions on the sale of cigarettes. Although this appeared to be a positive step because there were no more tobacco ads on television or the radio, there also were no more antitobacco ads. The rate of cigarette smoking began to rise again after these antismoking ads were pulled.

Significant Characters in Tobacco Advertising

Since there were no longer ads on television, tobacco advertising soon increased in other media. Increasingly, tobacco companies turned to new methods, or characters that consumers could identify with, to increase consumer desire for their products.

Joe Camel was a cartoon character developed in 1974 by the R.J. Reynolds Company and used to advertise Camel brand cigarettes. It was first used in a French ad campaign and made its U.S. debut in 1987. By 1991, the sale of Joe Camel merchandise, much of which appealed to kids, was earning R.J. Reynolds over $40 million yearly. In that same year, the *Journal of the American Medical Association* published studies which showed that Joe Camel was as well-known as Mickey Mouse among preschoolers, and that Camel's market share for smokers under age eighteen had skyrocketed—from under 1 percent in 1987 to almost a third of the market by 1991.

Clearly, cartoon characters worked well. In fact, one study tried using cartoon characters and a real person to deliver the same antismoking message. Young people remembered more details about the ad with the cartoon character, stated the ad was more believable, and were better able to comprehend and interpret the message.

Philip Morris created another highly successful character—the Marlboro Man. A rugged, individualistic cowboy, working hard with his horses on the range, invited smokers to "Come to Marlboro Country." They could leave behind all the hassles of the city or suburbia and escape with the Marlboro Man. Ironically, several

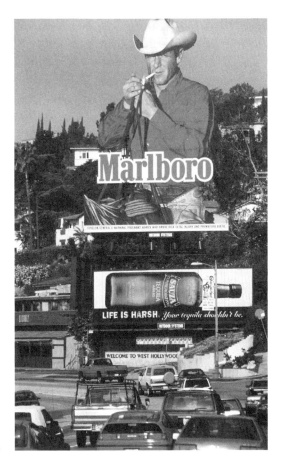

The Marlboro Man ad campaign employed by Philip Morris was a highly successful cigarette marketing strategy.

actors used in the Marlboro Men ads have since died of lung cancer including Marlboro Man Wayne McLaren. In the 1992 shareholder's meeting of Philip Morris McLaren informed the board that he had lung cancer as a result of his pack-and-a-half-a-day smoking habit and asked the company to reduce its advertising in the hopes of reducing the number of smokers. The company responded that they would not change its practices. Just four months later, McLaren was dead from lung cancer.

Tobacco in the Media

For years, many characters in movies and on television shows smoked. Thomas Stanislao, who smoked for twenty years before quitting, says, "I fell in love with old 1940s movies—what would Humphrey Bogart be if he weren't wreathed in romantic clouds of smoke? (Of course, it killed him at 58.) So I started to smoke. One minute I was a nonsmoker, the next I was a pack-a-dayer."[42]

Tobacco companies also advertise by paying large sums of money for brand-name or logo exposures in movies. In 1980, for about forty-two thousand dollars, Philip Morris purchased twenty-two exposures of the Marlboro logo in the movie *Superman II*. Lois Lane becomes a chain-smoker of Marlboro Lights. There are Marlboro billboards and a van with the Marlboro logo on it. The taxi in the final scene has a Marlboro sign on top. As a result, the company received much public exposure. Besides its initial run, *Superman II* was rerun on television countless times during the 1980s, is still available as a videotape rental, and still runs occasionally on syndicated television channels.

Tobacco companies used to have another way of getting media exposure. They used to be able to place ads around ballparks. When games were televised, the company's tobacco products were

A cartoon takes aim at the romantic image of the Marlboro Man.

instantly advertised. While those ads are now banned, tobacco companies can still get media exposure by sponsoring a vehicle in a sporting event. When they sponsor a car or boat in a race, that vehicle is often covered with the tobacco company's logo. Some boats have even been named after their sponsors, like the Winston Eagle and Miss Budweiser unlimited hydroplanes. Race car drivers and pit crew will wear uniforms emblazoned with the name or logo of the sponsor. During television or radio coverage, the brand is advertised constantly. However, recently there have been discussions about banning this practice.

In 2002, the Olympic Games were held in Salt Lake City, Utah. The Olympic Organizing Committee asked Olympic athletes to promote a healthy lifestyle and to promote sports as an alternative to tobacco. Utah law prohibits smoking in public places and the use of cigarettes by or the sale of cigarettes to those under the age of nineteen, one year older than in most states. Visitors to Utah were required to comply with Utah law even if they were considered of legal age to use tobacco in other states. There was a media campaign to increase the awareness of the Olympic tobacco policy and the Utah law among the many visitors from around the world, both athletes and spectators.

Today's Tobacco War: Waged in the Courts

The Olympic policy and the changing public opinion of tobacco use came as a result of hard-won legal battles. In the last five decades, since there have been studies and reports that increasingly verified the negative health effects of tobacco use, the tobacco industry has been embroiled in lawsuits. Many smokers filed suit, blaming the tobacco companies for their illnesses. Some of the lawsuits were successful. Others failed because the judges placed as much blame on the smokers for having decided to smoke as on the companies who produced these highly addictive products.

The tide began to turn in the 1990s because it was disclosed that tobacco companies knew about the detrimental health effects of nicotine all along. There were documents, memos, and other evidence that proved that the tobacco industry had known well

Tobacco executives testify before Congress in 1998. The tobacco companies had lied about their knowledge of nicotine addiction for decades.

before the 1964 surgeon general's report. In fact, the tobacco companies tried to reduce known carcinogens and develop more healthful cigarettes. Scientists also manipulated levels of nicotine and other chemicals in tobacco to maintain smokers' addiction. They clearly knew about nicotine's pharmacological actions.

However, the industry conspired to lie to the courts, Congress, and the American public. Tobacco executives testified before Congress that to the best of their knowledge, nicotine was not addictive, even though internal company memos and testimony of tobacco company whistleblowers (those who told the truth or "blew the whistle") stated the opposite. And yet, when confronted, the executives denied everything.

By unifying to deceive the public, the tobacco companies had removed the choice from the hands of consumers. Many consumers may have willfully chosen to smoke, but they were not aware, as the tobacco companies were, of the full health implications of tobacco usage or of the addictive nature of the product. When the extent of the deception was exposed, the courts began to favor the tobacco users who sued. By the mid-1990s, individual states initiated lawsuits against the tobacco companies seeking damages—primarily reimbursement for the costs that the states have paid to care for sick smokers.

The tobacco industry, facing the overwhelming evidence and testimony against them, finally admitted that it did indeed know of nicotine's addictive nature and of the health risks of tobacco use. The Liggett Group broke ranks from the other companies by agreeing to pay the Medicaid bills of California smokers in a lawsuit settled in 1996. In 1997, Liggett admitted it knew of the dangers of smoking for years. It settled with twenty-two states to pay for the health-care costs of smokers who were ill.

Lawsuits against other tobacco companies continued. In 1997, the states' attorneys general and the tobacco companies reached an agreement. More details were changed and worked out as the settlement passed through Congress in 1998. The result of the negotiations produced the Master Settlement Agreement of 1998 between the tobacco companies and the attorneys general of forty-six states and five U.S. territories.

In the settlement, the tobacco companies agreed to compensate the states for all the money that the states have spent to care for sick smokers. In addition, they agreed to run antismoking ads targeted toward young people and change their marketing practices so new ads are directed only toward adults. The tobacco companies must also develop research-based education programs designed to prevent and reduce youth smoking. The total tab for the tobacco companies will be more than $200 billion over twenty-five years.

The Price of Tobacco Today

The high costs of the 1998 settlement have directly raised the cost of tobacco products. Tobacco consumers are paying for a large chunk of the total amount of the tobacco companies' settlement. In addition to tobacco-company expenses like the settlement agreement, the inflated price of cigarettes and other tobacco products is a reflection of the government's belief that increasing the prices of cigarettes will discourage people from smoking as much.

Cash Crop

In the heat of the Great Depression, there were tough times for many smokers. Most could not afford cigarettes. Farmers of all kinds, including tobacco farmers, had crops that could not be sold, and there was soon a glut of tobacco in the marketplace. This led to government action. Congress passed the Agricultural Adjustment Act of 1933, which contained price supports for tobacco and other agricultural products. Farmers could continue to produce tobacco and be guaranteed a decent price for their crops.

The government also determined who could grow and sell tobacco. Those who grew tobacco were given quotas, which refer to burley tobacco, or allotments, which refer to other types of tobacco. The allotments or quotas entitled a farmer to produce one acre of tobacco or a certain number of pounds of tobacco. Quotas and allotments were initially established to help restrict tobacco production and make the prices steady. Tobacco farmers must buy or lease the allotments or quotas from the government. Some people who have inherited allotments or quotas lease their tobacco production rights to tobacco farmers.

Tobacco farmers are reluctant to give up tobacco growing because tobacco earns a higher price than other crops. After the 1998 settlement, the tobacco industry set aside money to help maintain the price farmers earn for tobacco.

Most tobacco farmers defend their right to produce and sell a legal product, especially one that earns so much money. For many farmers, tobacco money is the college fund for their children. Tobacco is also less labor-intensive than some other crops, requiring minimal care for a period of time, and then intense labor during harvesting. Some people have other jobs and grow tobacco for extra income and to ensure a comfortable style of living.

But some people, including some doctors who own allotments, choose not to use them because they no longer believe in selling tobacco. If a quota is not used over a period of years, it will expire and cannot be used again. Some allotments also expire; others are returned to a pool and can be reallocated. Some farmers have decided to change to other crops following the serious tobacco-related illness of a friend or family member.

Tobacco farmers pay a yearly assessment to ensure that the price of tobacco will be guaranteed, even in off years. With that assessment and the funds that tobacco companies set aside to help farmers, tobacco growing is a fantastic cash crop. It will continue to be a good business for farmers as long as there are tobacco users.

The government taxes tobacco products at a higher rate than other products because of the costs that states and the federal government have incurred to treat sick smokers and address public-health issues related to tobacco use. Each state decides how much to tax, what to do with the tax revenues, and also what to do with the tobacco settlement money. There is much debate about the disbursement of these funds.

Minnesota state senate majority leader Roger Moe describes a project in which schoolchildren made a memorial wall with photos, drawings, and stories of relatives who died or who were ill due to tobacco use:

> The hundreds of pictures of dead moms and dads, dead grandparents, dead aunts and dead uncles help focus the public policy question of what to do with the [tobacco settlement and tax] money. . . . We get so bogged down talking about dollar amounts and endowments, and this makes it all real. I think I don't know of any more graphic way you could show the cost of smoking to Minnesota.[43]

Diversification

As more and more people recognize the immense human costs of tobacco use, the number of tobacco users has plummeted. Yet tobacco companies face financial difficulties when tobacco users quit in record numbers or when existing users die and new smokers are not there to maintain a base number of tobacco users. The industry has realized this for years and as a result has moved toward diversification of their product offerings. They have purchased other consumer-oriented and food businesses, acquiring, over the years, companies such as Burma Shave, Seven-Up, Kraft, Nabisco, General Foods, Jell-O, and Maxwell House Coffee. By adding new businesses, tobacco companies do not have to depend on changing consumer beliefs about tobacco or laws governing tobacco usage and sales to make a profit.

However, some people feel it is socially responsible to boycott products sponsored by tobacco money and will not purchase or use those products. In June 2001, Philip Morris offered the Kraft Food Company as an initial public offering—essentially its own company with common shares traded on the stock market. However, Philip

Morris retained most of the preferred stock, thus keeping 98 percent of the voting rights for the company. Some people still boycott Kraft foods because a tobacco company is the primary owner.

Nicotine as a Pharmaceutical

One controversial business sector that tobacco companies are entering into is the health-care industry. Whether people like it or not, the tobacco companies are experts in the field of nicotine use. For years, they have studied nicotine and its effects on the body. So, as they think of new business prospects, it makes sense for tobacco companies to use their knowledge for health-care purposes. In 1999, R.J. Reynolds launched a new company called Targacept, which focuses on selling Reynolds's expertise to help develop nicotinic drugs.

Nicotine has been described as a means for patients to medicate themselves. Certain groups of people, such as those with schizophrenia, almost universally smoke. Many patients with bipolar disorder smoke. Although it is not certain why, nicotine appears to help people with some mental illnesses to function better.

Even though tobacco use is unhealthy, studies are investigating whether smaller doses of nicotine in a safer delivery system (a patch or an inhaler, for example) can provide medical benefits in some cases. Certain diseases like Parkinson's disease and Alzheimer's disease have been associated with a dopamine deficit. Because nicotine increases dopamine production, studies have shown it may help the functioning and memory of some of these patients.

In other studies, patients with Down syndrome also experienced improvements in functioning with nicotine use. Tourette's syndrome was also improved by nicotine, although most patients had serious side effects since many of the patients in the clinical trials were children. Ulcerative colitis also showed improvement with nicotine, although scientists are not sure why. Other studies have shown that while high doses of nicotine impair blood-vessel tone, low doses can actually promote growth of blood vessels. More research is needed to see if blood vessel growth promoted by nicotine can have a therapeutic effect in the body.

Some people think it is ethically wrong for the tobacco companies to profit from their nicotine expertise—expertise that has been won at the cost of illness and death to hundreds of thousands of people; expertise that has been used to manipulate the tobacco product; and expertise that was used to deceive and deliberately lie to the American public and the tobacco consumer for so many years. Others think it is commendable that the tobacco companies are using their knowledge in a way that is helpful instead of harmful.

The Future for Big Tobacco

Besides therapeutic uses, tobacco companies are focusing attention on other open markets. Many companies realize the prospects for future tobacco sales in the United States are diminishing and the market as they knew it has been altered irrevocably. In the

A man purchases cigarettes in Tajikistan. American tobacco companies are increasingly focusing on profits to be made in foreign markets.

United States, the antismoking campaign and advertisements, the public-health messages, and the shift in social attitudes are turning the tide against tobacco. Although there will probably always be tobacco users, the number of tobacco smokers in the United States is dropping, and it may continue to decline as more smokers heed the health warnings.

However, this is not the situation in other countries. Many developing countries are far behind the United States in terms of public-health awareness and product regulations. American cigarettes are being exported in ever-increasing numbers. In many countries, tobacco regulations are not as strict, and U.S. tobacco companies can market to adolescents as well as adults. Teens are just as vulnerable to advertising and promotional items in other countries as they are in the United States.

Back on the home front, lawsuits abound, and there are serious financial repercussions for tobacco companies as a result of all the legal action. In the 1998 settlement, the state governments held the tobacco industry accountable for illnesses and increased costs of medical treatments related to tobacco use. The U.S. federal government initiated a lawsuit in 1999 for its losses in providing Medicare and other assistance to tobacco users. The case is still pending. Settlement talks were initiated in June 2001, although the government is still pursuing litigation as well as a settlement.

Other countries have filed suit against U.S. tobacco companies in the U.S. court system as well. Venezuela and Bolivia filed lawsuits in 1999. Russia filed suit in 2000; Tajikistan and Kyrgyzstan in 2001. The European Union and countries such as Ecuador have also filed suits. Most lawsuits claimed that the tobacco companies deliberately harmed their citizens, deceived them by not telling the truth about tobacco and nicotine, and caused astronomical medical costs. Some of these lawsuits have been dismissed or settled, and some have yet to be decided.

In addition to these lawsuits, U.S. states are watching the tobacco industry carefully to make sure it is complying with the 1998 tobacco settlement agreement. In March 2001, five states sued R.J. Reynolds, stating that they were not keeping to the terms of the agreement. There were still concerns about advertising, promotional items displaying tobacco company logos, and other issues. Individual smokers continue to sue tobacco companies for damages. The continuing lawsuits will affect the ongoing viability of individual tobacco companies and the tobacco industry overall.

At one time over half of the men and more than a third of the women in the United States smoked. There has been substantial progress since that time. Yet the use of nicotine products is still pervasive in our society. In 1998, 47.2 million Americans were smokers while 148.6 million Americans were former and nonsmokers. If tobacco was eliminated, over 430,000 deaths each year could be prevented and over $97.2 billion could be saved in medical costs and lost productivity.

Only time will tell whether tobacco will continue to be a viable product in today's business climate. Prohibition or outlawing of both alcohol and tobacco occurred in the early part of the twentieth century. It did not work—people went to clandestine hideaways called speakeasies and continued to smoke and drink despite the penalties for noncompliance. Now, as long as there are tobacco users and tobacco is legal, there will continue to be a demand for tobacco products.

Notes

Introduction: Nicotine's Iron Grip

1. Jerry Thomas, interview by author, Pitkin, LA, December 22, 1999.
2. Bennett LeBow, "Attorney General's Settlement Agreement," CNN, March 20, 1997. http://europe.cnn.com/US/9705/tobacco/docs/liggett.html.

Chapter 1: Tobacco's Roots in the Western World

3. Robert Burton, *The Anatomy of Melancholy*, 1621. Reprint, New York: Tudor, 1948.
4. Quoted in Richard Kluger, *Ashes to Ashes: America's Hundred-Year Cigarette War, the Public Health, and the Unabashed Triumph of Philip Morris*. New York: Alfred A. Knopf, 1996, p. 14.
5. Kluger, *Ashes to Ashes*, p. 14.
6. Kluger, *Ashes to Ashes*, p. 113.
7. Brief History of Tobacco Advertising to Women, "1928 Lucky Strike—Reach for a Lucky Instead of a Sweet." http:/speakerskit.chestnet.org/04/ppt_pages/a_set/a_03.htm.
8. Brief History of Tobacco Advertising to Women, "1928 Lucky Strike—Reach for a Lucky Instead of a Sweet." http:/speakerskit.chestnet.org/04/ppt_pages/a_set/a_03.htm.

Chapter 2: Addicted: In the Lungs and in the Brain

9. David Sanford, interview by author, e-mail, May 11, 2001.
10. Stephen King, "10 Pages a Day," *Writer's Digest*, April 2001, p. 34.

11. Steven Hoadley, interview by author, e-mail, July 24, 2001.
12. Elizabeth Whelan, "Editorial: Many Factors Play a Role in Cigarette Smoking Addiction," drkoop.com, September 21, 2000. www.drkoop.com/news/focus/september/smoke_addiction.html.
13. Christina Johnson, interview by author, e-mail, March 26, 2001.
14. Manda Djinn, interview by author, e-mail, March 25, 2001.
15. Quoted in BBC News, "Cigarettes 'Cut Life by 11 Minutes,'" December 31, 1999. http://news.bbc.co.uk/hi/english/health/newsid_583000/583722.stm.
16. Susannah Tate, interview by author, e-mail, April 10, 2001.
17. Norah Troy Teeter, interview by author, Mountlake Terrace, WA, July 29, 2001.

Chapter 3: Teens, Tobacco, and Trade-Offs

18. Mona Vanek, interview by author, e-mail, March 24, 2001.
19. Barb Chandler, interview by author, e-mail, March 24, 2001.
20. Becky Brooks, interview by author, e-mail, March 24, 2001.
21. Thomas Stanislao, interview by author, e-mail, March 24, 2001.
22. Quoted in Philip J. Hilts, *Smokescreen: The Truth Behind the Tobacco Industry Cover-Up.* Reading, MA: Addison-Wesley, 1996, p. 98.
23. Laurie P. Teeter, interview by author, Tacoma, WA, August 12, 2001.
24. Brooks, interview.
25. Quoted in *Nando Times,* "Economists Say Cigarette Tax Will Cut Teen Smoking," May 26, 1998. http://archive.nandotimes.com/newsroom/ntn/biz/052698/biz7_23306_noframes.htm.
26. Amy Bloxham, interview by author, Tacoma, WA, June 25, 2001.
27. Myra Nelson, interview by author, e-mail, March 25, 2001.
28. Jennifer Nelson, interview by author, e-mail, March 26, 2001.

Chapter 4: Nixing Nicotine

29. Madeleine Armstrong, interview by author, e-mail, March 24, 2001.
30. Kathleen Purcell, interview by author, e-mail, March 24, 2001.
31. Jodi Waxman, interview by author, e-mail, January 25, 2001.
32. Colleen Brady, "Ask an Expert: Questions and Answers on Health and Family," *Chatelaine*, November 2000, p. 22.
33. Judith Stock, interview by author, e-mail, March 25, 2001.
34. Dave Waddle, interview by author, Mountlake Terrace, WA, August 14, 2001.
35. Frederick Nehser, interview by author, Woodinville, WA, December 10, 2001.
36. Jennifer Nelson, interview.
37. Quoted in S. Avery, "No Magic in Kicking the Habit: Success Is Found in Combination of Therapies," *Raleigh News and Observer*, June 4, 2001.
38. Stanislao, interview.

Chapter 5: Advertising, Big Business, and Uncle Sam

39. Quoted in Borio, "The History of Tobacco Part III."
40. Quoted in Borio, "The History of Tobacco Part III."
41. Quoted in Borio, "The History of Tobacco Part III."
42. Stanislao, interview.
43. Quoted in David Hanners, "Remembering Lives up in Smoke: Rally Marks Settlement; Prevention Efforts Urged," *Pioneer Press*, May 8, 1999.

Organizations to Contact

Action on Smoking and Health
(ASH) 2013 H St., NW, Washington, DC 20006
(202) 659-4310
website: http://ash.org

A national, legal-action antismoking organization that supports and protects rights of nonsmokers.

American Cancer Society (ACS)
1599 Clifton Rd., NE, Atlanta, GA 30329
(800) 227-2345
website: www.cancer.org

A nationwide organization dedicated to eliminating cancer by sponsoring research and education and promoting advocacy and service.

American Lung Association (ALA)
1740 Broadway, New York, NY 10019
(800) Lung-USA • (212) 315-8700
website: www.lungusa.org

Although interested in the prevention of all lung disease, today's ALA places special focus on asthma, tobacco control, and maintaining a lung-healthy environment.

Americans for Nonsmokers' Rights (ANR)
2530 San Pablo Ave., Suite J, Berkeley, CA 94702

(510) 841-3032
website: www.no-smoke.org

This national lobbying organization is dedicated to nonsmokers' rights. By lobbying against tobacco at the local and national level, ANR seeks to promote laws that help protect nonsmokers from environmental tobacco smoke (secondhand smoke).

COST: Children Opposed to Smoking Tobacco
e-mail: costkids@costkids.org
website: www.costkids.org

This organization was started by middle-school students who wanted to keep smoking and tobacco products away from kids.

National Center for Tobacco-Free Kids
1400 Eye St., Suite 1200 Washington, DC 20005
(202) 296-5469
website: www.tobaccofreekids.org

This organization's purpose is to protect children from becoming addicted to tobacco, to reduce or eliminate kids' exposure to secondhand smoke, and to alter the public's acceptance of tobacco.

National Clearinghouse for Alcohol and Drug Information (NCADI)
PO Box 2345 Rockville, MD 20847-2345
(800) 729-6686 • TTY: (800) 487-4899
website: www.health.org

A service of the Substance Abuse and Mental Health Services Administration, NCADI offers a number of free or low-cost printed resources on a number of topics including alcohol, drug abuse, and abuse prevention.

National Spit Tobacco Education Program (NSTEP)
410 North Michigan Ave., Chicago, IL 60611
(312) 836-9900
website: www.nstep.org/nstep.htm

NSTEP seeks to break the connection between baseball and smokeless tobacco and to educate the public about this dangerous and even deadly drug.

Nicotine Anonymous
419 Main St., PMB #370, Huntington Beach, CA 92648
(866) 536-4539 • fax: (714) 969-4493
website: www.nicotine-anonymous.org

Nicotine Anonymous uses a twelve-step approach to help participants lead nicotine-free lives. Participants share experiences, strength, and hope with each other.

Office on Smoking and Health (OSH)
Centers for Disease Control and Prevention
Mailstop K50 4770 Buford Highway NE, Atlanta, GA 30041
(800) 232-1311 • (770) 488-5705
website: www.cdc.gov/tobacco

This office works toward preventing tobacco use by young people, encouraging quitting, and protecting nonsmokers from environmental tobacco smoke. The OSH website provides full text of various surgeon general's reports, including the landmark 1964 Surgeon General's Report on Smoking and Health.

STAT (Stop Teenage Addiction to Tobacco)
Northeastern University
360 Huntington Ave., 241 Cushing Hall, Boston, MA 02115
(617) 373-7828 • fax: (617) 369-0130
website: www.stat.org

A national organization whose goal is to stop childhood and teenage tobacco dependence by partnering with young people to promote activism, provide education, and encourage advocacy and networking.

Tobacco BBS
PO Box 359, Village Station, New York, NY 10014-0359

(212) 982-4645
website: www.tobacco.org

Gene Borio, the master of this site, lets the media speak for itself. He scans a variety of news sources regularly, and posts quotes about tobacco current events. This site has free resources as well. It contains information on tobacco news, research, history, control, cessation (quitting), and much more.

U.S. Food and Drug Administration
5600 Fishers Ln., Rockville, MD 20857-0001
(888) FDA-4KIDS
website: www.fda.gov

This agency regulates the manufacture and sale of food and pharmaceuticals. The toll-free number is available for information or to report the sale of tobacco to a minor.

Youth Media Network (YMN)
17872 Moro Rd., Prunedale, CA 93907
(831) 663-9208 • In California: (800) 733-8377
website: www.ymn.org

A project that is funded by a tax on tobacco, this nonprofit, California-based group encourages young people to create tobacco-free messages, poems, and short stories. YMN also recognizes and spotlights young people and community or school groups who are exemplary in tobacco-use prevention and education.

For Further Reading

Gina De Angelis, *Nicotine and Cigarettes*. Philadelphia: Chelsea House, 2000. Provides a multifaceted look at nicotine, including well-researched sections on tobacco's lengthy history, what nicotine does in the body, and why people use tobacco.

Arlene Hirschfelder, *Kick Butts! A Kid's Action Guide to a Tobacco-Free America*. Parsippany, NJ: Julian Messner, 1998. This upbeat and well-written book provides young people with information about being and becoming tobacco free.

Elizabeth Keyishian, *Everything You Need to Know About Smoking*. Rev. ed. New York: Rosen, 2000. Includes information on health risks, why people smoke, the choices people make regarding smoking, addiction, how to quit, and how to get help.

Barbara Moe, *Teen Smoking and Tobacco Use: A Hot Issue*. Berkeley Heights, NJ: Enslow, 2000. This book answers questions about teen smoking—who smokes, what kinds of tobacco they use, when they start, where they use it, why they stay hooked, how to resist, and how to quit. Includes numerous quotes and firsthand accounts by teen smokers.

Mary E. Williams, ed., *Smoking*. San Diego, CA: Greenhaven Press, 2000. This book presents a number of essays and articles, each of which examines opposing viewpoints about smoking-related issues.

———, *Teen Smoking*. San Diego, CA: Greenhaven Press, 2000. A collection of essays and articles that examines the issue of teen smoking. Articles address teen tobacco use, race issues, advertising, peer pressure, use of tobacco in media and movies, and more.

Works Consulted

Books

Robert Burton, *The Anatomy of Melancholy,* 1621. Reprint, New York: Tudor, 1948. A collection of insights and information about physical and psychological health in the seventeenth century.

Philip J. Hilts, *Smokescreen: The Truth Behind the Tobacco Industry Cover-Up.* Reading, MA: Addison-Wesley, 1996. This book is an exposé of the tobacco industry's deceptive practices. It includes information about illegal tobacco marketing to youth, legal skirmishes, and government interventions (or lack thereof).

Institute of Medicine, *Growing Up Tobacco Free: Preventing Nicotine Addiction in Children and Youths.* Washington, DC: National Academy Press, 1994. This fascinating 250-page report is the culmination of an eighteen-month study commissioned to examine the prevention of nicotine dependence among children and young people.

Richard Kluger, *Ashes to Ashes: America's Hundred-Year Cigarette War, the Public Health, and the Unabashed Triumph of Philip Morris.* New York: Alfred A. Knopf, 1996. This comprehensive Pulitzer prize–winning book looks at the history of tobacco use through the ages, describing in riveting detail how cigarettes came to be one of the United States's premier products for individual coping and self-medicating.

Periodicals

Alcoholism and Drug Abuse Weekly, "Habits, Attitudes Vary Among Young Smokers," July 24, 2000.

———, "Studies Find Significant Impact From Pro- and Anti-Smoking Ads," March 6, 2000.

———, "Survey: Adolescent Smoking Linked to Risky Behaviors," July 3, 2000.

Wael K. Al-Delaimy, Julian Crane, and A. Woodward, "Passive Smoking in Children: Effect of Avoidance Strategies at Home as Measured by Hair Nicotine Levels," *Archives of Environmental Health*, March 2001.

American Medical News, "Surprise: Nicotine May Actually Be Beneficial Sometimes," March 20, 2000.

———, "Tobacco Sickness Hits 41% of Farm Workers," March 6, 2000.

Anil Ananthaswamy, "Nicotine's Fatal Attraction: Nicotine Raises Dopamine in Brains of Rats," *New Scientist*, August 26, 2000.

S. Avery, "No Magic in Kicking the Habit: Success Is Found in Combination of Therapies," *Raleigh News and Observer*, June 4, 2001.

Ursula E. Bauer, Tammie M. Johnson, Richard S. Hopkins, and Robert G. Brooks, "Changes in Youth Cigarette Use and Intentions Following Implementation of a Tobacco Control Program: Findings From the Florida Youth Tobacco Survey, 1998–2000," *Journal of the American Medical Association*, August 9, 2000.

Body Bulletin, "Do You Smoke to Cope?" February 2000.

———, "What's in Cigarette Smoke?" April 2000.

Bruce Bower, "Forbidden Flavors: Scientists Consider How Disgusting Tastes Can Linger Surreptitiously in Memory," *Science News*, March 29, 1997.

Colleen Brady, "Ask an Expert: Questions and Answers on Health and Family," *Chatelaine*, November 2000.

Jennifer Braunschweiger, "The Unfiltered Truth: Why Cigarettes Are Deadlier for Women," *Reader's Digest*, July 2001.

Paul H. Brodish, "The 'Smoking' Unborn," *Priorities for Health*, March 31, 1999.

Brown University Child and Adolescent Behavior Letter, "Cartoons and Tobacco: Using Cartoon Characters in Commercials to Warn Children Against Smoking," February 2001.

———, "Teenagers Susceptible to Tobacco Marketing," April 2001.

Brown University Digest of Addiction Theory and Application, "Survey Indicates Increase in High School Seniors' Smoking Habits," November 1999.

Maria Chang, "Hooked on Nicotine," *Science World*, December 14, 1998.

Pamela I. Clark, Sharon L. Natanblut, Carol L. Schmitt, Charles Wolters, and Ronaldo Iachan, "Factors Associated with Tobacco Sales to Minors: Lessons Learned From the FDA Compliance Checks," *Journal of the American Medical Association*, August 9, 2000.

Andy Coghlan, "Deadly Draw," *New Scientist,* October 7, 2000.

Consultant, "Twentieth-Century Efforts to Extinguish Tobacco Use," April 15, 2001.

Contemporary Pediatrics, "Teen Tobacco Use High and Varied," April 2000.

W. Marvin Davis, "Tobacco/Nicotine Dependence and Cessation Therapies," *Drug Topics,* September 7, 1998.

Drug Topics, "Does Nicotine Have a Role in the Treatment of Ulcerative Colitis?" July 7, 1997.

Economist, "Great Drug, Shame About the Delivery System," September 23, 2000.

Kathleen Fackelmann, "Smokers' Hearts Don't Pick Up Pace," *Science News,* August 30, 1997.

Family Practice News, "Beedie Babies," May 15, 1999.

Arthur J. Farkas, Elizabeth A. Gilpin, Martha M. White, and John P. Pierce, "Association Between Household and Workplace Smoking Restrictions and Adolescent Smoking," *Journal of the American Medical Association,* August 9, 2000.

Mary Beth Flanders-Stephans and Sara G. Fuller, "Physiological Effects of Infant Exposure to Environmental Tobacco Smoke: A Passive Observation Study," *Journal of Perinatal Education,* March 31, 1999.

Ernest H. Friedman, Geoffrey C. Williams, and Edward L. Deci, "Presentation Style as Important as the Message," *Archives of Pediatrics and Adolescent Medicine,* March 2000.

Elizabeth Goodman and John Capitman, "Depressive Symptoms and Cigarette Smoking Among Teens," *Pediatrics,* October 2000.

David Hanners, "Remembering Lives up in Smoke: Rally Marks Settlement; Prevention Efforts Urged," *Pioneer Press,* May 8, 1999.

Health and Medicine Week, "Teens with Emotional and Behavioral Problems More Likely to Smoke," March 5, 2001.

Health Letter on the CDC, "Nicotine Can Improve Some Aspects of Cognitive Performance," March 6, 2000.

Nicole Hendricks, "A New Leaf: Former Tobacco Farmers Find a New Crop," *Country Journal,* September–October 1998.

Agneta Hjalmarson, Fredrik Nilsson, Lars Sjostrom, and Olle Wiklund, "The Nicotine Inhaler in Smoking Cessation," *Archives of Internal Medicine,* August 11, 1997.

Kimberly A. Horn, Xin Gao, Geri A. Dino, and Sachin Kamal-Bahl, "Determinants of Youth Tobacco Use in West Virginia: A Comparison of

Smoking and Smokeless Tobacco Use," *American Journal of Drug and Alcohol Abuse,* February 2000.

Alice T.D. Hughes, Elizabeth Goodman, and John Capitman, "Depressive Symptoms and Cigarette Smoking Among Teens," *Journal of the American Medical Association,* December 20, 2000.

J.R. Hughes, M.G. Goldstein, R.D. Hurt, and S. Shiffman, "Recent Advances in the Pharmacotherapy of Smoking," *Journal of the American Medical Association,* January 6, 1999.

Richard D. Hurt, "Nicotine Patch Therapy in 101 Adolescent Smokers: Efficacy, Withdrawal Symptom Relief, and Carbon Monoxide and Plasma Cotinine Levels," *Journal of the American Medical Association,* April 12, 2000.

Journal of the American Medical Association, "Bidi Use Among Urban Youth—Massachusetts, March–April 1999," October 20, 1999.

———, "Trends in Cigarette Smoking Among High School Students—United States, 1991–1999," September 27, 2000.

Christine Kilgore, "Teen Antismoking Program Sparks Decline in Use," *Family Practice News,* May 15, 2000.

Stephen King, "10 Pages a Day," *Writer's Digest,* April 2001.

David Lazarus, "Up in Smoke: Even Lower Taxes Can't Save San Francisco's Struggling Cigar Shops," *San Francisco Chronicle,* July 1, 2001.

Joseph K. Lim, "Tackling Tobacco in the Twenty-First Century," *Journal of the American Medical Association,* April 26, 2000.

Maclean's, "Big Tobacco Concedes," February 9, 1998.

Howard Markel, "Behavior: Ask the Expert—Teen Smoking," *Contemporary Pediatrics,* January 2001.

Maurice W. Martin, Sarah Levin, and Ruth Saunders, "The Association Between Severity of Sanction[s] Imposed for Violation of Tobacco Policy and High School Dropout Rates," *Journal of School Health,* October 2000.

Leah Martino, "Kaiser Smoking Innovations Program: Clinical Trial to Study the Effectiveness of Acupuncture for Smoking Cessation," *California Journal of Oriental Medicine,* April 30, 1999.

Mike Mitka, "Surgeon General's Newest Report on Tobacco," *Journal of the American Medical Association,* September 20, 2000.

Eric T. Moolchan, "Is It a Good Time for Treatment? Smoking Prevention for Teenagers," *Brown University Child and Adolescent Behavior Letter,* January 2001.

Morbidity and Mortality Weekly Report, "Tobacco Use Among Middle and High School Students—United States, 1999," January 28, 2000.

Kelly Morris, "Nicotine Withdrawal Impairs Attention," *Lancet,* April 24, 1999.

Sonia Nichols, "Nicotine By-Product Distresses Vessel Tone in Animal Model," *Angiogenesis Weekly,* February 9, 2001.

Pain and Central Nervous System Week, "Smoking May Increase Risk (of Developing Agoraphobia or Other Panic Disorders)," December 9, 2000.

S. Roan, "Struggle to Quit," *Los Angeles Times,* February 14, 2000.

Science News, "Hooked on a Feeling: Nicotine Addiction," June 19, 1999.

———, "Nicotine Addiction Curbed by New Drug," January 2, 1999.

Rainer Seidl, Monika Tiefenthaler, Erwin Hauser, and Gert Lubec, "Effects of Transdermal Nicotine on Cognitive Performance in Down's Syndrome," *Lancet,* October 21, 2000.

Lorena M. Siqueira, Linda M. Rolnitzky, and Vaughn I. Rickert, "Smoking Cessation in Adolescents: The Role of Nicotine Dependence, Stress, and Coping Methods," *Archives of Pediatrics and Adolescent Medicine,* April 2001.

John Travis, "Nicotine Metabolism May Spawn Carcinogen," *Science News,* October 28, 2000.

Bill Tuttle, "My War with a Smoke-Free Killer," *Reader's Digest,* October 1996.

Vaccine Weekly, "Nicotine Vaccine Shows Promise for Combating Tobacco Addiction," January 12, 2000.

Melanie A. Wakefield, Frank J. Chaloupka, Nancy J. Kaufman, C. Tracy Orleans, Dianne C. Barker, and Erin E. Ruel, "Effect of Restrictions on Smoking at Home, at School, and in Public Places on Teenage Smoking: Cross-Sectional Study," *British Medical Journal,* August 5, 2000.

Glenn D. Walters, "Spontaneous Remission from Alcohol, Tobacco, and Other Drug Abuse: Seeking Quantitative Answers to Qualitative Questions," *American Journal of Drug and Alcohol Abuse,* August 2000.

Karl L. Yen, Elizabeth Hechavarria, and Susan B. Bostwick, "Bidi Cigarettes: An Emerging Threat to Adolescent Health," *Archives of Pediatrics and Adolescent Medicine,* December 2000.

Todd Zwillich, "Hypocritical Oath: R.J. Reynolds Launches Targacept, a Company That Will Sell RJR's Nicotine Expertise to Companies

Developing Nicotinic Drugs Used Therapeutically," *Family Practice News,* August 1, 1999.

Internet Sources

William A. Alcott, *The Use of Tobacco: Its Physical, Intellectual, and Moral Effects on the Human System.* Rev. ed. New York: Fowler and Wells, 1876. www.members.tripod.com/medicolegal/alcott1836.htm.

BBC News, "Cigarettes 'Cut Life by 11 Minutes,'" December 31, 1999. http://news.bbc.co.uk/hi/english/health/newsid_583000/583722.stm.

Gene Borio, "A Few of Our Losses from Smoking," QuitSmokingSupport.com, 1995–2001. www.quitsmokingsupport.com/noteables.htm.

———, "The History of Tobacco Part III: Twentieth Century—The Rise of the Cigarette," History On-Line, 1997. www.historian.org/bysubject/tobacco3.htm.

———, "Tobacco Timeline," Tobacco BBS, 1993–2001. www.tobacco.org/History/Tobacco_History.html.

David M. Bresnahan, "Baseball and Tobacco: A Dying Tradition," InvestigativeJournal.com, April 20, 1998. www.investigativejournal.com/chewball.htm.

Brief History of Tobacco Advertising to Women, "1928 Lucky Strike—Reach for a Lucky Instead of a Sweet." http://speakerskit.chestnet.org/04/ppt_pages/a_set/a_03.htm.

Christine Gregoire, "Order to Cease and Desist Internet Sales of Cigarettes to Minors," Office of the Attorney General (WA), December 20, 1999. www.wa.gov/ago/pubs/bidis_order.htm.

Thomas Hariot, "A Briefe and True Report of the New Found Land of Virginia", 1588. www.people.virginia.edu/~msk5d/hariot/main.html.

Bill Koenig, "Profile: A Painful Portrait," *USA Today,* June 6, 1996. www.usatoday.com/sports/baseball/sbbw0429.htm.

David Lazarus, "Cigarette Prices Up Again: Cost of Pack Rising Around 22 Cents," *San Francisco Chronicle,* August 31, 1999. www.sfgate.com/cgi-bin/article.cgi?file=/chronicle/archive/1999/08/31/BU29950.DTL.

Bennett LeBow, "Attorney General's Settlement Agreement," CNN, March 20, 1997. http://europe.cnn.com/US/9705/tobacco/docs/liggett.html.

Ann Meeker-O'Connell, "How Nicotine Works," HowStuffWorks, 1998–2001. www.howstuffworks.com/nicotine4.htm.

Nando Times, "Economists Say Cigarette Tax Will Cut Teen Smoking," May 26, 1998. http://archive.nandotimes.com/newsroom/ntn/biz/052698/biz7_23306_noframes.htm.

Office on Smoking and Health, "1988 Surgeon General's Report: The Health Consequences of Smoking: Nicotine Addiction." www.cdc. gov/tobacco/sgr_1988.htm.

Christine H. Rowley, "What Happens to Your Body After Your Last Cigarette?" Smoking Cessation on About.com, 2001. http://quitsmoking.about. com/library/bllastoc.htm.

Spit Tobacco Prevention Network, "The Faces of Spit Tobacco," 1996–2001. www.flash.net/~stopn/Boys.html.

Stonee's Native American Lore, Legend and Teachings, "How They Brought Back Tobacco," 1996–1997. www.ilhawaii.net/~stony/ 697myths.html.

Virtual Office of the Surgeon General, "You Can Quit Smoking: A 5-Day Plan to Get Ready," March 2001. www.surgeongeneral.gov/tobacco/ 1stweek.htm.

———, "You Can Quit Smoking: Tips for the First Week," March 2001. www.surgeongeneral.gov/tobacco/1stweek.htm.

Elizabeth Whelan, "Editorial: Many Factors Play a Role in Cigarette Smoking Addiction," drkoop.com, September 21, 2000. www.drkoop. com/news/focus/september/smoke_addiction.html.

WhyQuit.Com, "Nineteen-Year-Old Sean Marsee's Tobacco Message," 1999–2001. www.whyquit.com/whyquit/SeanMarsee.html.

John B. Wight, *Tobacco: Its Use and Abuse.* Columbia, SC: L.L. Pickett, 1889. www.members.tripod.com/medicolegal/wight1889.htm.

Index

Index

Index 111

R.J. Reynolds Company, 51, 81, 89, 91–92
Roanoke Island, 15
Rolfe, James, 18
Roosevelt, Franklin Delano, 23
Russell, Rosalind, 76

sailors, 21
Sanford, David, 31
San Salvador, 12
scarves, 26
schizophrenia, 89
schools, smoking prevention programs in, 55–56
secondhand smoke. *See* passive smoking
Seven-Up, 88
shaman, 14
Siberia, 18
silk rectangles, 26
smokeless tobacco, 8, 21, 29–30, 38, 45
smoking
    ages when people start, 9
    banned at 2002 Olympic Games, 84
    exercise and, 36
    first Europeans who became smokers, 12
    health hazards of, 8, 27, 39–42
    laws passed to limit or prohibit, 17–18, 54–55
    methods of, 7–8
    number of smokers in United States, 10, 23
    passive, 42–43
    prevention programs in schools, 55–56
    reasons people smoke, 8–9
Smoking and Health (report), 27
smoking-cessation groups, 66–68, 79
snakebites, 14
snuff, 8, 21, 29–30, 57
speakeasies, 92
spirits, 14
spittoons, 21
Stanislao, Thomas, 50, 72, 82–83
Stanwyck, Barbara, 76
stillbirths, 41
Stock, Judith, 63–64
Stonee's Native American Lore, Legends, and Teachings (website), 13
stress, 47–50
stroke, 39
Students Working Against Tobacco

(SWAT), 56–58
sudden infant death syndrome, 41
Superman II (film), 83
Supreme Court, 79
surgeon general, 10, 27, 73, 76–79
SWAT. *See* Students Working Against Tobacco
Swift, Eleanor, 36

Tainos, 12
Tajikistan, 91
Talihina High School, 57
tar, 35, 39, 75
Targacept, 89
Tate, Susannah, 40
taxes, 17–18, 54, 88
teenagers
    marketing tobacco to, 50–53
    number of, who smoke, 45
    reasons tobacco is used by, 46–50, 64
    smoking prevention programs for, 54–58
    stress-copying tips for, 48
Teeter, Laurie, 54
Teeter, Norah, 43
Terry, Luther, 76
tobaca. See pipes
tobacco
    advertising of, 25–27, 50–53, 75–76, 79–84
    allotments for farmers of, 87
    banning of, 92
    Columbus receives as gift, 12
    Congress requires warning labels for, 79
    described, 6
    English grow in Virginia, 18–20
    exporting to foreign countries, 91
    future of, 90–92
    health benefits of, 78
    health hazards
        chemicals found in, 35
        from chewing tobacco, 8, 21
        damage to body from, 37–40
        deaths from, 17, 39–40
        from passive smoking, 42–43
        suits filed against tobacco companies because of, 84–86, 91–92
        surgeon general's report on, 27, 76–79
        in teenagers, 53–54
        for women, 40–42

About the Author

Jenny Rackley is a freelance writer living in the Seattle area. Besides writing, she is a licensed teacher and a mother of three smoke-free children. Ms. Rackley graduated from Wellesley College with a degree in computer science and has since earned an MBA. She has been a speaker at several writers' conferences and has published over 150 articles for parents, teachers, and young people. The majority of Ms. Rackley's publications have focused on parenting, nutrition, and health-related topics.